THE CHALLENGE IN MATHEMATICS AND SCIENCE EDUCATION

PSYCHOLOGY'S RESPONSE

THE CHALLENGE IN MATHEMATICS AND SCIENCE EDUCATION

PSYCHOLOGY'S RESPONSE

Edited by
Louis A. Penner
George M. Batsche
Howard M. Knoff
Douglas L. Nelson

American Psychological Association, Washington, DC

Published by
American Psychological Association
750 First Street, NE
Washington, DC 20002

Copies may be ordered from
APA Order Department
P.O. Box 2710
Hyattsville, MD 20784

In the UK and Europe, copies may be ordered from
American Psychological Association
3 Henrietta Street
Covent Garden, London
WC2E 8LU England

Typeset in Century Book by Techna Type, Inc., York, PA

Printer: Braun-Brumfield, Inc., Ann Arbor, MI
Cover Designer: Michael David Brown, Inc., Rockville, MD
Technical/Production Editor: Miria Liliana Riahi

Library of Congress Cataloging-in-Publication Data
The Challenge in mathematics and science education : psychology's response / edited by
    Louis A. Penner ... [et al.].
        p.   cm.
    Includes bibliographical references and index.
    ISBN 1-55798-207-4
    1. Mathematics—Study and teaching—United States.   2. Science—Study and teach-
ing—United States.   3. Minority students—Education—United States.   I. Penner, Louis
A., 1943–   .
QA13.C43   1994   93-21336
370.15′65—dc20   CIP

British Library Cataloguing-in-Publication Data
A CIP record is available from the British Library.

Printed in the United States of America
First Edition

APA Science Volumes

A PA expects to publish volumes on the following conference topics:

Changing Ecological Approaches to Development: Organism–Environment Mutualities

Converging Operations in the Study of Visual Selective Attention

Emotion and Culture

International Conference on the Psychology of Industrial Relations

Maintaining and Promoting Integrity in Behavioral Science Research

Measuring Changes in Patients Following Psychological and Pharmacological Interventions

Sleep Onset: Normal and Abnormal Processes

Stereotypes: Brain–Behavior Relationships

Temperament: Individual Differences in Biology and Behavior

Women's Psychological and Physical Health

As part of its continuing and expanding commitment to enhance the dissemination of scientific psychological knowledge, the Science Directorate of the APA established a Scientific Conferences Program. A series of volumes resulting from these conferences is jointly produced by the Science Directorate and the Office of Communications. A call for proposals is issued several times annually by the Science Directorate, which, collaboratively with the APA Board of Scientific Affairs, evaluates the proposals and selects several conferences for funding. This important effort has resulted in an exceptional series of meetings and scholarly volumes, each of which individually has contributed to the dissemination of research and dialogue in these topical areas.

The APA Science Directorate's conferences funding program has supported 28 conferences since its inception in 1988. To date, 18 volumes resulting from conferences have been published.

William C. Howell, PhD
Executive Director

Virginia E. Holt
Assistant Executive Director

Contents

Contributors

Harry P. Bahrick, Ohio Wesleyan University

George M. Batsche, University of South Florida, Tampa

Andrew S. Becker, University of New Mexico

John Bransford and the Cognition and Technology Group, Vanderbilt University

Henry C. Ellis, University of New Mexico

James J. Gallagher, University of North Carolina

James G. Greeno, Stanford University

Janet Shibley Hyde, University of Wisconsin

James M. Jones, American Psychological Association

Daniel P. Keating, Ontario Institute for Studies in Education

Howard M. Knoff, University of South Florida, Tampa

Barbara L. McCombs, Mid-continent Educational Laboratory

Diane McGuinness, University of South Florida, Ft. Myers

Douglas L. Nelson, University of South Florida, Tampa

Louis A. Penner, University of South Florida, Tampa

Robert S. Siegler, Carnegie Mellon University

Charles D. Spielberger, University of South Florida, Tampa

Larry J. Varner, University of New Mexico

Foreword

Education in America is broadly viewed as a system in crisis. Helping students to develop competence in mathematics and science is of particular concern because of predicted catastrophic shortfalls by the turn of the century with regard to the number of scientists, engineers, and technicians who will be needed to "run the country" and keep America economically competitive. What can be done to help American children acquire the knowledge and skills needed to respond to the challenges of an ever-changing society in communities beset by drugs, disrupted family structure, childhood abuse and neglect, and poverty and homelessness?

In its early years, many of the leaders of the American Psychological Association (APA)—notably past presidents William James and John Dewey—devoted a major part of their energies to enhancing American education. James's informal talks with teachers are among his most interesting and exciting contributions to the literature in psychology. Dewey's emphasis on the needs and interests of the learner continue to dominate current efforts to achieve educational reform.

Interest in learning has been a central theme in psychology since the establishment of this discipline and has generated extensive research. Although much of this work has focused on investigations of memory, cognition, and the learning process, there has been relatively little concern about applying psychological knowledge to improve American education. Over the past decade, however, an increasing number of psychologists have directed their attention to the psychological processes involved in learning mathematics and to other factors that might improve educational achievement. Unfortunately, relatively few applications of this research have bridged the gap between the laboratory and actual school settings. To do so will require psychologists to work directly with educators and colleagues in other scientific disciplines to develop valid and useful applications of psychological knowledge for improving curricula and instruction and for the enhancement of learning in school and community settings.

During my tenure as APA president, I endeavored to foster and stimulate a greater commitment of the energy and resources of our Association to improving American education. At the 1991 APA Annual Convention in San Francisco, the President's Psychology in Education (PsyEd) Task Force organized a mini-convention that focused on contributions of psychology to learning and education. The APA Board of Educational Affairs joined with nine APA divisions in contributing convention time to an exciting program of invited addresses and symposia that encompassed 20 hours of stimulating presentations. More than 60 participants examined topics relating to three major themes: (a) cognition, affect, and individuality in learning and education; (b) contributions of developmental, cognitive, and social psychology to shaping education reform; and (c) psychology in the school of the future.

In 1991, the APA Board of Scientific Affairs also took an important step to support applications of psychology to education. Funds were allocated for a national conference to identify advances in psychological research with the potential for enhancing educational outcomes. At this conference, titled "Contributions of Psychology to Education in Science and Mathematics," participants examined basic and applied research relevant to applications of developmental, cognitive, affective, and motivational processes to the learning and teaching of mathematics and science. This conference served as a springboard to the focused collection of chapters that now constitute this book.

The conference was organized by Professors Louis Penner and Douglas Nelson of the Psychology Department at the University of South Florida (USF) and by Professors Howard Knoff and George Batsche, Directors of the School Psychology Program in the USF College of Education. Colleagues invited to participate in the conference were selected on the basis of their expertise in psychological science as well as on that of their sensitivity to the application of this knowledge in educational settings. Much of the credit for convening the conference and for publishing this volume is due to the leadership, organizational skills, and persistence of

Professor Penner, who served as chair of the Organizing Committee that contributed substantially to the conference itself, to preparing the introductions for each section of this book, and to editing the chapters in their areas of expertise.

We sincerely hope that this volume will contribute to further efforts to advance American education, the APA, and the discipline of psychology. Other activities stimulated and sponsored by the PsyEd Task Force include a second national conference, "Assessment of Learning and Educational Achievement," convened in collaboration with the Mid-continent Regional Educational Laboratory and held in November 1991 at The Johnson Foundation Wingspread Conference Center in Racine, Wisconsin. A third national conference, "Research to Practice: Improving the Learning and Teaching of Science and Mathematics," was organized by APA in collaboration with the Council of Scientific Society Presidents, the National Science Teachers Association, the National Council of Teachers of Mathematics, the National Association of Biology Teachers, and the American Chemical Society. Supported by a grant from the National Science Foundation, this conference was sponsored by the Council of Scientific Society Presidents and was held in September 1992, also at the Wingspread Conference Center in Racine.

A major priority of the PsyEd Task Force was to develop, within the APA, an organization for high-school psychology teachers. We estimate that more than 8,000 teachers are currently involved in providing some form of instruction in psychology for almost one million high school students, and these numbers are increasing. A new organization for high school psychology teachers has recently been established by the APA Board of Educational Affairs, adopting as its name, Teachers of Psychology in Secondary Schools (TOPSS). Through TOPSS, which was approved by the APA Council of Representatives in February 1993, the APA can now be more effective in helping high school psychology teachers to improve their instruction by keeping them informed of current research findings. APA also plans to stimulate and facilitate the development of constructive links between high school teachers and college and university psychology departments.

Another exciting project initiated by the PsyEd Task Force is the development of a high school psychology textbook. The APA Office of Communications is presently exploring the feasibility and desirability of moving forward with this project, which was stimulated by a highly successful program undertaken by the American Chemical Society in developing and publishing a high school textbook, *ChemCom* (Chemistry in the Community). Published in 1986, *ChemCom* presents theoretical principles and research applications of chemistry in the context of a student's everyday experience. With substantial grant support from the National Science Foundation, physicists and biologists are presently developing books modeled after *ChemCom*, *Active Physics*, and *BioCom*. These efforts also provide excellent models for the APA in developing a high school textbook that will focus on psychological science and important applications of psychology in community settings.

Education is and should be a unifying force within psychology. The education and training of psychologists is unique among the mental health professions in providing a strong scientific foundation for practitioners and a continuing commitment to serving the public interest. It is time for the APA and the discipline of psychology to make the contributions to American education that it is uniquely qualified to make.

Charles D. Spielberger
President, 1991–1992
American Psychological Association

Preface

This book has a history that is similar to many academic efforts. It began with informal conversations in a mail room, lay dormant for many months because of other commitments, died and was later resurrected several times, and was finally brought to completion in a series of marathon work sessions. More specifically, this effort began several years ago when then American Psychological Association (APA) President-Elect Charles Spielberger informally suggested that the University of South Florida (USF) host a conference on psychology and education. A working committee was established and from this larger group came the people who turned this general idea into a proposal and obtained the funds to hold the conference that led to this book.

The editors of this book have backgrounds in several areas of psychology. One is a social psychologist (Penner), another a cognitive psychologist (Nelson), and two are trained as school psychologists (Batsche and Knoff). Despite this diversity (at least in terms of psychologists), we share two common perspectives. First, we feel that the educational problems regarding mathematics and science education are real, important, and demanding of our attention. Second, we believe that psychology has much to contribute to the solution of these problems. Thus, we unite in this book to describe and analyze some of the contributions of psychology to mathematics and science education. Within this broad topic, we particularly wish to emphasize issues related to individuals who have historically been "at-risk" with regard to mathematics and science achievement—namely, members of ethnic minorities and women.

We wish to express our gratitude for grants from the Science Directorate of APA and the Office of Sponsored Research at the USF. Additional support was provided by William G. Katzenmeyer, Dean of the College of Education, and Gerhard G. Meisels, Provost, at the USF. Without their support and encouragement, we would not have been able to present this book.

We are also grateful to Lewis Lipsitt and Virginia Holt of the APA Science Directorate, to Rollin C. Richmond, Dean of the USF College of Arts and Science, and to George R. Newkome, Vice President for Sponsored Research at USF, for their substantial contributions. Debbie Touchton did a remarkable job in her capacity as conference coordinator. She handled everything with incredible skill and with as much calm as could be reasonably expected under the circumstances. We are also extremely appreciative of the efforts of B. J. Thomas, the office manager of the USF Psychology Department.

Finally, we express our thanks and deep gratitude to our authors. They gave their time and their effort in exchange for very little by way of rewards. They met our deadlines and requests for changes with few complaints. Most important, they agreed to share their expertise and wisdom with us and with you. I hope that this will be a major gift to the discipline and to the task of educating the children of North America.

Louis A. Penner
George M. Batsche
Howard M. Knoff
Douglas L. Nelson

Introduction

Louis A. Penner

R ecent widespread public concern over the crisis in mathematics and science education can be traced to the 1989 National Educational Summit held in Charlottesville, Virginia. At this meeting, former President George Bush and the governors of all 50 states (including current President Bill Clinton, who was governor of Arkansas at the time) presented a set of goals for America's primary and secondary educational system. These goals stressed the need for the following: to develop a plan for the education of minority and at-risk students; to focus on technological training, with an emphasis on mathematics and science achievement; and to create an educational system that is accountable to the people who support it and has an impact on those who use it. The summit resulted in a formal plan to address the nation's educational needs. This plan, *America 2000* (United States Department of Education, April 1991), established six specific national education goals that were to be achieved by the year 2000. (A complete list of these goals may be found in chapter

1 in this volume.) Of most relevance to this book are two goals that specifically concern mathematics and science education. The first of these is that, by the year 2000, all students will leave high school with mastery over "challenging subject matter, including English, Mathematics, Science, History, and Geography." The second goal is that, by the year 2000, students in the United States will rank first in the world in mathematics and science education. To many people these goals might seem quite modest. Indeed, we suspect that many Americans believe that, by the time they have graduated from high school, students must have achieved mastery over the basic principles of mathematics and science. Perhaps even more people believe that the United States actually leads the world in science and mathematics education.

These beliefs are at odds with the realities of mathematics and science education in the United States. The data from national surveys of academic performance of American students and comparisons of their mathematics and science skills with those of students from other countries yield a very alarming picture of education in these fields in the United States. Because these data are presented in detail in several chapters in this book, we highlight only some of the findings that caused the president, the governors, and most educators such concern.

Have American students achieved mastery over basic aspects of mathematics and science by the time they graduate from high school? Less than 10% of American high school graduates have any degree of specialized knowledge in science; indeed, most high school graduates do not even have the requisite basic skills in mathematics or science to enable them to be productive members of the work force without substantial additional training. In 1992, almost 40% of the American schoolchildren who took the test developed by the National Assessment Governing Board failed to reach basic proficiency levels in mathematics. Only 2% of the 12th graders reached advanced levels in mathematics (Associated Press, January 14, 1993).

Is American mathematics and science education at the primary and secondary levels superior to that of other industrialized nations? American students consistently score far below students from other countries on standardized tests of mathematics and science skills. For example, the mathematical achievement of the typical American eighth-grader ranks 12th among the 18 leading industrialized nations. The typical American ninth-grader ranks next to last in science achievement when compared with the ninth-graders in these other countries, and American high school students' performance in advanced science classes, such as biology, is last among the nations examined.

Thus, the appropriate question about the goals contained in *America 2000* (1991) is not, "Have we not already achieved them?" Rather, it is, "Can they ever be achieved?" Most educators agree that, even under the best of circumstances, the task is daunting. Most also agree that these are not the best of times for the American educational system. There is a continuing decline in the willingness of the public to pay for the education of American children (or, at least, the willingness of the politicians who purport to represent them). Even if this decline in resources could be reversed, however, there is reason to believe that, in the future, it will be even more difficult to move substantially ahead in mathematics and science education.

To understand the reason why this appears to be the case, we must consider the expected changes in the demographics of the school-age population and the data on how these demographic variables are related to academic achievement. In the next 20 to 25 years, the percentage of schoolchildren who are both White and male will shrink relative to other segments of the school-age population. By the year 2020, minority children (primarily African American and Hispanic), who currently constitute less than 30% of the U.S. school-age population, will be at least 45%. There is no reason to believe that the overall percentage of men in the school-age population will decline markedly from what it is now, but more of them will be poor and from minority populations.

Second, for a variety of reasons (many of which are discussed in this book), students who are White and male do much better, on average, on standardized tests of mathematics and science achievement than do minority group and female students. For example, the total Scholastic Aptitude Test (SAT) scores for White students are 27% higher than those of African–American students and about 15% to 20% higher than those of Hispanic students. With regard to gender, men consistently outperform women in mathematics. In the 1992 National Assessment of Educational Progress, male 12th graders, on average, did much better in mathematics than did female 12th graders. Furthermore, gender differences are most pronounced among the students of greatest ability.

Therefore, in the future there will be a smaller percentage of students who come from the group that currently is most likely to score the highest on standardized tests of achievement in mathematics and science. It must be reemphasized, moreover, that even these students are not performing that well when compared with international norms. Thus, we must be concerned both with improving the performance of our best students in mathematics and science and with finding ways to improve the performance of students who have traditionally been at risk in these areas.

Efforts to improve the mathematics and science performance of members of ethnic minorities and of women could probably be justified solely on the basis of the principles and values that are part of this country's political, ethical, and moral heritage. Even if one rejected this political argument for programs that benefit these groups, however, enlightened self-interest would lead to the same course of action. James Gallagher (in chapter 1) makes this argument quite cogently, noting that, in the 21st century in the United States, there will be an even greater dependence on science and technology to maintain a reasonable standard of living and economic viability than there is today. We are, however, facing a dramatic shortage of science and engineering students in the United States. Massey (1989) predicted that by the year 2006, the demand for scientists and engineers will outstrip the supply by about 400,000 people.

It was against the background of these problems that *America 2000* (1991) was developed. The plan contains some interesting and some controversial proposals for improving the American educational system. Among them are the following: the creation of public and private partnerships to fund innovative new school programs; the establishment of schools in each congressional district that would be specifically designed to achieve the goals of the plan; allowing parents to choose which schools their children would attend (including private schools); and the establishment of national standards and testing for educational achievement.

As is often true with proposals that deal with important public policy, *America 2000* (1991) evoked both praise and criticism from the people who read it (see, for example, the November 1991 *Phi Delta Kappan*). Some of these negative reactions seem to be reasonable. For example, it is difficult to understand how giving parents the freedom to choose the schools that their children would attend would improve the quality of mathematics and science education (especially for economically disadvantaged ethnic minorities). It is similarly difficult to comprehend why the plan gave little attention to the role of social and economic inequalities as causes of the educational problems of many schoolchildren, or why the plan simply ignored the question of how the major changes that it advocated would be financed. Despite its shortcomings, however, there was widespread acceptance of most of the goals of *America 2000*.

From the perspective of contemporary American psychology, there is another problem with the recommendations emanating from both the Charlottesville Education Summit and *America 2000*: Neither addressed any of the *mechanisms* that would produce the recommended results. That is, largely missing in either document is any discussion of the underlying scientific principles that must necessarily drive the process of educational change. It was this deficiency, or shortcoming, in these plans to reform and to improve the American educational system, that led to our decision to hold the conference from which the chapters of this book evolved.

Psychology's Contribution to Mathematics and Science Education

The idea that psychology and psychological principles can be used to improve the educational process is not a new one. Consider, for example, the widespread adoption of B. F. Skinner's ideas and techniques in the American educational system (Lattal, 1992). It would be incorrect, however, to believe that the involvement of psychology in education is only 40 or 50 years old. Indeed, psychologists' interest in the educational process predates even the formal founding of psychology as a scientific discipline. Hergenhahn (1992) identified Johann Friedrich Herbart, an early 19th-century philosopher, as the first "educational psychologist." What is striking about Herbart's views on the best ways to educate children is how similar they are to many of the ideas presented in this book. For example, Herbart strongly believed that one must consider the learner's own experiences and perspective if one is to maximize his or her learning. This advice seems entirely consistent with the "learner-centered psychological principles for enhancing education" that are discussed in chapter 9 in this book and that are being developed by the American Psychological Association's Task Force on Psychology in Education (McCombs, 1992).

Of course, Herbart was not the only philosopher and psychologist to offer contemporary educational advice. Jean-Jacques Rousseau (1762/ 1947) told the educators of his time, "take time to observe nature; watch your scholar well before you say a word to him; first leave the germ of his character free to show itself, do not constrain him from anything, the better to see him as he really is" (p. 58). As one reads chapters 5, 10, and 11 in this volume, one will see that Rousseau's advice is still being followed by psychologists interested in the educational process.

So, psychology's interest in and involvement with educational change and reform is not new. What is (relatively) new is the impact of the cognitive revolution on education and vice versa. As observed in chapter 5, cognitive psychology can contribute to the "basic assumptions that underlie the organization of our school curriculum, our assessments of student achievement, and important aspects of teachers' classrooms practices, especially in mathematics and science." At the same time, research in educational settings can add to "the fundamental knowledge, theory, and scientific practices of the study of situated cognition and learning."

Organization of This Book

The chapters in this book represent a set of contemporary perspectives on how to improve mathematics and science education in the United States. They are divided into the following four parts, each with an introduction: educational context; cognitive, affective, and situational factors in mathematics and science education; social and cultural factors in mathematics and science education; and applied psychological research on mathematics and science education.

Given the earlier discussion of psychology's contributions to educational practices, it is not surprising that this volume reflects the growing communality between the interests of cognitive psychologists and educators concerned with changing and improving current educational practices. The book is not solely cognitive in its orientation, however. In several of the chapters, the reader will find the application of principles from social psychology, sociology, and even anthropology to the problems that exist in mathematics and science education. Quite reasonably, moreover, several of the chapters draw on more than one psychological perspective in their attempts either to understand the learning process or to propose ways in which it can be improved.

In passing, it should be noted that the authors are not all of one opinion about the causes of the problems discussed earlier or their solutions. This diversity of views may disturb some people. We see it, however, as a reflection of the complexity of the issues that confront us and of the unwillingness of the authors to accept simple solutions to difficult and demanding problems. We would propose that this is how good applied science progresses.

Part 1 features a chapter by James Gallagher, in which he cogently and succinctly presents the problems that confront those who wish to improve the quality of mathematics and science education in the United States. However, in contrast to most of the other contributors in this volume, who concentrate on children who are at risk with regard to mathematics and science achievement, Gallagher focuses primarily on effectively educating the more talented mathematics and science students.

Gallagher's chapter is opinionated and provocative. His views will, no doubt, annoy some people and anger others as he argues against what he sees as the unwarranted intrusion of certain political considerations into the educational process. The chapter is, however, an appropriate way in which to begin this book. However provocative they are, Gallagher's opinions are thoughtful and informed. Moreover, as his title suggests, one cannot really separate educational practice from public policy and polit-

ical philosophy. Gallagher is probably right when he proposes that sometimes what is politically "good" or "correct" may not be educationally
sound and vice versa. This is perhaps a message that should be repeated
throughout the book and remembered by those who attempt to make the
changes that are needed.

Part 2 contains contributions by four distinguished cognitive psychologists. What is striking to me, as a social psychologist, is how applied
these chapters are. Indeed, in some regards, they may be the most applied
chapters in the book. This was inconsistent with many of my own stereotypes about the interests of the "typical" cognitive psychologist. Of
course, some of this emphasis simply reflects the kind of chapters that
we asked the authors to write, but the chapters also reflect a real connection between cognitive and educational psychology.

For example, Harry Bahrick's chapter reports on his research on
the long-term retention of academic material and offers specific recommendations for increasing the amount of course content students actually
retain. In the following chapter, Henry Ellis and his associates address
the role of emotional factors in the learning process and demonstrate the
importance of considering both cognition and emotion in understanding
the learning process. They draw from the research literature in emotion
and cognitive processes to suggest interventions that may improve children's ability to learn and retain new material. In the next chapter, John
Bransford presents the extremely exciting work of the Cognition and
Technology Transfer Group at Vanderbilt University. Using the knowledge
base created by cognitive psychologists and state-of-the-art video technology, this team of educators and scientists has developed new and
innovative ways to teach young children about the basic principles of
mathematics and science. And in the final chapter of Part 2, James Greeno
draws on concepts from cognitive and social psychology to develop a
theory of learning that could be applied to curriculum development, as
well as to instructional methods.

Part 3 explicitly addresses the variables of ethnicity and gender as
they relate to academic performance in mathematics and science. As
discussed earlier, these demographic variables are most relevant to any

consideration of the problems confronting mathematics and science education in the United States.

In the first chapter in this section, James Jones discusses possible causes of academic difficulties for students from ethnic minorities and suggests some ways to address these problems. In doing this, Jones provides persuasive evidence of the importance of cultural variables in educational successes and failures. The next two chapters concern gender as an important variable in the education process and provide an illustration of the diversity of viewpoints and opinions noted earlier. In her chapter on gender differences in mathematics achievement, Janet Shibley Hyde uses the results of meta-analyses to argue that there are few, if any, innate differences between males and females in mathematical abilities. In the following chapter, however, Diane McGuiness takes issue with this position and argues for a biological rather than a sociological interpretation of differences in the performance of girls and boys in mathematics and science. As noted earlier, we are not discouraged by the fact that such controversies still exist. Both points of view are well supported and presented, and it is likely that the eventual resolution of this theoretical disagreement will incorporate some of *both* of the positions represented in these chapters.

Part 4 provides an exemplar of how applied psychological science can facilitate educational change. The initial chapter, by Barbara McCombs, presents a specific proposal for improving the way that we teach our children. Much of this proposal is based on the results of research efforts such as those represented in this book. The final two chapters concern research programs that specifically address learning difficulties that affect educational achievement in mathematics and science. Daniel Keating discusses how individual differences ("developmental diversity") in cognitive and social development might affect achievement in mathematics and science. Robert Siegler describes his research on the strategies that low-income children use to solve mathematics problems. Both chapters suggest ways in which instructional methods and educational curricula might be designed so as to maximize their effectiveness with at-risk children.

Although we have edited the chapters, we have done little to change the tone, or "voice," of the authors. These are professionals who have learned much about education and have much to teach us. We believed it best that they speak for themselves. We hope that psychologists and educators alike will carefully read each of the chapters, learn from them, and take what they have learned to the classroom where the children wait for their help.

The Time Is Now

Phrases such as *the crisis in education* and *the critical problems facing educators* are liberally sprinkled throughout this book. We suspect, however, that many people may have difficulty in really appreciating the importance of the particular crisis in mathematics and science education in the United States. They may have become desensitized to such alarmist phrases in the media, or they simply may not see this particular problem as any more of a threat to our nation's future than any of the other "national crises" about which they have read recently. To be sure, the crisis in mathematics and science education is not particularly dramatic or vivid. It is unlikely to garner large headlines or to be the subject of the lead story on the evening news, but our failure to educate our children adequately in these disciplines does demand our attention. If we are unable to change and improve the way that our schools teach these subject areas, there will be unfortunate, but very likely, consequences.

First, although the absolute number of children in the schools will increase, the percentage of them who choose careers in mathematics- and science-related fields will decline. As suggested earlier, this will be seen most clearly among members of ethnic minorities and women, resulting in serious shortages in scientists and engineers. This would be an intolerable waste of the talent and potential of many women and members of ethnic minorities.

Second, because our schools are not adequately training students in the basics of mathematics and science, the private and public sectors of our economy will have to devote more resources to postsecondary or remedial training of their personnel just so that these employees will be able to do their jobs. This situation will create a tremendous additional burden on our economy and limit our economic growth. Finally, in the 21st century, the nation will either be forced to import the technology that it needs or to find itself falling behind the other major industrialized nations in this regard.

The major consequence of these outcomes is that the United States will become increasingly unable to meet the new challenges and crises that will inevitably confront it in the 21st century. There will be many tragedies associated with these new and unforeseen problems, but perhaps the greatest tragedy will be that, in the 20th century, we had the tools and the technology to prevent many of them, yet we did nothing. Thus, the time to change the way that we educate our children is now. It is our hope that this book will contribute to this process.

References

America 2000. (1991, November). *Phi Delta Kappan, 73*, 176–253.

Hergenhahn, R. W. (1992). *An introduction to the history of psychology*. Belmont, CA: Wadsworth.

Lattal, K. A. (1992). Introduction to special issue on B. F. Skinner and psychology. *American Psychologist, 47*, 1269–1271.

Massey, W. E. (1989, September). Science education in the United States: What the scientific community can do. *Association Affairs*, 915–921.

McCombs, B. L. (1992). *Learner centered psychological principles: Guidelines for school redesign and reform*. Washington, DC: American Psychological Association.

Rousseau, J. J. (1947). *The social contract* (C. Frankel, Trans.). New York: Macmillan. (Original work published 1762)

U.S. Department of Education. (1991). *America 2000: An education strategy*. Washington, DC: Author.

U.S. students lag in education. (1993, January 14). Washington, DC: Associated Press.

PART ONE

The Educational Context

An Intersection of Public Policy and Social Science: Gifted Students and Education in Mathematics and Science

James J. Gallagher

How does important knowledge find its way from the laboratories and scholarly halls of academia to the front line of educational practice? We know that the process is nonlinear and even haphazard. One of the themes of this chapter is that, in addition to the usual communication problems, there may be strong societal resistance to ideas that do not fit into current value orientations. This is a likely occurrence when considering the education of gifted students. They often seem to be caught in a debate about equity and excellence in American education.

Education, like medicine, agriculture, and engineering, is an application field that is dependent, in large measure, on information derived from more basic sciences. Education, like medicine, has developed research models and a knowledge base in its own right, particularly on such subjects as the process of educational practice. The contributions of psychology and the social sciences to the practice of education have been many and impressive, from comprehending the intellectual performance of children to understanding the affective life of students, to the organization of knowledge in various curricula, and to those complex social processes that so complicate schools and concern school administrators and teachers.

If the contributions of social science have been less satisfying or extensive than might have been hoped, perhaps that is because there is no simple translation process by which knowledge is moved from the laboratory or the research journal to the complex community known as the public schools. Like crude oil from the ground, raw knowledge is valuable, but it becomes truly useful only after being transformed and synthesized into consumer products. Scientific information is our intellectual crude oil, and we have not put many resources into a systematic approach to translation and knowledge synthesis.

This chapter deals with a specific area of interest to educators and, intermittently, to the general public—the development of outstanding student mathematicians and scientists. A full understanding of these issues involves much more than the study of superior cognitive processes, complex as these are (Feldman, 1986). If we see the student as the *figure* in a *figure–ground* model, then the *ground*, the ecological setting in which the education of such students takes place, can be important to the understanding of gifted students. This fuller understanding may require input from sociology, social psychology, and organizational psychology, among others.

The Nature of Intelligence

Many distinguished psychologists have contributed to the general understanding of intelligence and its development. This chapter focuses on one dimension of this issue: the area of individual differences of those students who learn much faster than others. Terman and Oden (1947) and Hollingworth (1942) were some of the earliest scientists to pursue knowledge about gifted students and creativity. Their longitudinal studies revealed both high achievement in adulthood and substantial underachievement by a subset of the group. One clear bit of evidence on how difficult it is to translate research into educational policy is the inflexible rules on school entry dates and on the organization of grades by age level. Both these policies are contradictory to our knowledge of individual differences.

In the past decade, a series of investigators interested in the nature of intelligence and information processing has emerged. Simon (1980), Sternberg and Subart (1991), H. Gardner (1985), Borkowski and Kurtz (1987), and many others have introduced concepts—such as *metathinking* and *executive processes*—to the standard array of intellectual processes—*memory, classification, reasoning,* and *evaluation*—that have been familiar to educators for some time. The consequences of this new work are beginning to be evident only now, and little current impact on education has been noticed beyond a mere awareness that such work is underway.

For example, much of the identification of gifted students still uses tests based on the *g* factor, although H. Gardner's (1985) seven intelligences have begun to receive attention. The research reported on gifted students, as a consequence, will be on students who have performed well on the instruments that have been in use for some time.

Whereas Terman and Oden (1947) believed in the *g* factor and in the strong genetic influences on intelligence (after all, he called his work *The Genetic Studies of Genius*), the current child development position is that there is a complex and sequential interaction pattern between genetics and environment that tends to progressively facilitate or inhibit the full development of youngsters with special talents (Sameroff, 1986). As noted in Table 1, the development of symbolic systems such as language lie at the heart of more sophisticated intellectual development. Children who have been raised in an atmosphere in which language is not used extensively, or in which the parent is not present to interact with the child, will probably result in the child's limited language development, which, in turn, will lead to less than full-potential academic performance and to a possible consequent lack of interest in school and school-related activities. The combination of these progressive interactions, then, could result in a lower score on intelligence or aptitude measures than would have been likely under more optimum conditions.

Just as a series of unfavorable environmental forces can result in less favorable educational and psychometric outcomes, so can the op-

TABLE 1
Gardner's Multiple Intelligences

Linguistic intelligence	The ability to use language in written and oral expression to aid in remembering, solving problems, and seeking new answers to old problems (novelist, lecturer, lawyer, lyricist).
Logical–mathematical intelligence	The ability to use notation and calculation to aid intelligence with deductive and inductive reasoning (mathematician, physicist).
Spatial intelligence	The ability to use notation and spatial configurations. Important in pattern recognition (architect, navigator, sculptor, mechanic).
Bodily–kinesthetic intelligence	The ability to use all or part of one's body to perform a task or fashion a product (dancer, athlete, surgeon).
Musical intelligence	The ability to include pitch discrimination; sensitivity to rhythm, texture, and timbre; the ability to hear themes; production of music through performance or composition (musician).
Interpersonal intelligence	The ability to understand other individuals (their actions and motivations) and to act productively on that knowledge (teacher, therapist, politician, salesperson).
Intrapersonal intelligence	The ability to understand one's own feelings and motivations, cognitive strengths, and styles (just about anything).

Note. Adapted from "Giftedness from a multiple intelligences perspective" by V. Ramos-Ford and H. Gardner, 1991, pp. 56–58, in *Handbook of Gifted Education*, edited by N. Colangelo and G. A. Davis, Needham Heights, MA: Allyn & Bacon. Copyright © 1991 by Allyn and Bacon. Adapted by permission.

posite be true. If the family is encouraging and supportive, if the learning environment is superior, then there may be an opportunity for students from particular families and cultural subgroups to show a greater than average prevalence of high ability or aptitude (Bloom, 1985; Olszewski, Kulieke, & Buscher, 1987). Higher than expected prevalence of being identified as gifted has been found in children from Asian families. The high prevalence of Asian–American children in programs for gifted students, as well as in other areas of performance such as music and the arts, has been a reminder of the attention paid, in many Asian–American families, to the importance of education and of setting high expectations for children's performance.

The new emphasis on information processing has spawned some instructive and educationally relevant findings. One fruitful line of investigation has been a series of studies comparing the thinking processes of expert versus novice practitioners in fields such as medical diagnosis, physics problem solving, and chess playing (Rabinowitz & Glaser, 1985). In all of these investigations, a different problem-solving style was noted between expert and novice. The expert was able to refer to and use a rich network of associations that had been established with experience and time, whereas the novice practitioner was most often bound by the parameters of the particular problem to be solved. This distinction between expert and novice appears to have a direct analogy to that between gifted learners and average learners.

The obvious educational implication is that the student should be encouraged to build these rich bodies of associations within and across various content fields and asked to draw on these network associations to solve problems. Otherwise, the student may be building what Whitehead (1929) referred to as *inert knowledge*, that is, factual information clearly present in memory banks, but not usable for specific problems. For many people, geometric theorems and logarithms are clear examples of inert knowledge.

American Students and Progress

The apparent limited ability of American students to master mathematical and scientific concepts has been of general educational concern for at least three decades, starting with a serious educational reform effort after the sputnik scare (Bruner, 1960; Goodlad, 1964) and made manifest by major curriculum reform efforts initiated by the National Science Foundation and, later, by the U.S. Office of Education through projects such as the School Mathematics Study Group, the Chemical Bond Approach, the Biological Sciences Curriculum Study, and many others.

Current concerns in mathematics and science are reflected in the six National Education Goals adopted by the 50 governors and by President Bush in 1989 and now incorporated into *America 2000*, an effort to bring these goals to fruition. The goal that we "will be first in the world in mathematics and science by the year 2000" may seem puzzling. Are we not first now? The answer to that question is a major and painful story.

There are three sources of information to be brought to bear on this question: international comparisons, by which students from other countries are compared with our own on common measures of mathematical and scientific proficiency; performance on the National Assessment of Educational Progress (NAEP), a major 20-year effort to find out what American students know in various subject areas; and some individual, cross-cultural studies.

International and NAEP Comparisons

During the past few years, a series of reports has compared students from various countries on educational concept mastery in areas of mathematics and science. Most of these comparisons have shown students in the United States to be at a substantial disadvantage when compared with students from other countries, particularly from those on the Pacific Rim.

Among some of the recent findings (International Association for the Evaluation of Educational Achievement, 1988) have been the following:

- At Grade 5, the United States ranked in the middle in science achievement relative to 14 other participating countries.
- At Grade 9, U.S. students ranked next to last.
- In the upper grades of secondary school, "advanced science students" in the United States ranked last in biology and performed behind students from most countries in chemistry and physics.

Harold Stevenson et al. (1985) completed a study comparing Japanese, Chinese, Canadian, and U.S. students and arrived at similar, even startling, conclusions. For example, he found that the average math performance of primary-grade Chinese children falls at the 97 percentile of American students.

Stevenson et al. reported that American students are superior in only one dimension. They believe that they are doing well in mathematics and so do their parents. The Chinese students are less confident of their ability, although they outscore the American and Canadian children by a large margin.

Another index of academic performance is the NAEP, which provides an opportunity to judge the performance of students against what teachers believe they should have mastered at the age levels of 9, 13, and 17 years. In terms of an NAEP report card (Mullis & Jenkins, 1988), however,

- More than half of the nation's 17-year-olds appear to be inadequately prepared either to perform competently jobs that require technical skills or to benefit substantially from specialized on-the-job training. The thinking skills and science knowledge possessed by these high school students also seem to be inadequate for informed participation in the nation's civic affairs.
- Only 7% of the nation's 17-year-olds have the prerequisite knowledge and skills thought to be needed to perform well in college-level science courses. Because high school science proficiency is a good predictor of whether a young person will elect to pursue postsecondary studies in science, the probability that many more students will embark on future careers in science is very low.

The report concluded the following:

Students' knowledge of science, and their ability to use what they know, appear remarkably limited. That a very small proportion of junior high school students and only about 40% of high school students can be considered even moderately versed in this subject area is a matter of grave concern, as is the very small percentage (7%) of high school students with any degree of specialized knowledge in science. (Mullis & Jenkins, 1988, p. 61)

When results are broken down by gender and by race (Black and Hispanic students), the picture becomes even more discouraging. Although science cannot be reserved for a small elite group, even the best of students seem to be performing at only a mediocre level compared with reasonable expectations.

Shortage of Students in Science

Unfortunately, as reliance on science and technology has grown, the number of American scientists has declined. Already, professional scientists project a shortage of 400,000 scientists and engineers by the year 2006, the year in which most children born in 1984 would receive their undergraduate degrees from college (Massey, 1989). With the continuing shortage of scientists and with students' manifest lack of interest in the sciences, a serious challenge exists to current American superiority in these fields.

Jones (1988) reported that only a minority of PhD graduates in science and engineering from U.S. universities were native born. Science has not been seen as attractive, particularly to women and minority groups. In 1990, Blacks made up 12% of the nation's population, and only 2% of its scientists and engineers; Hispanics made up 9% of the nation's population and 2% of its scientists and engineers; and women made up 51% of the nation's population, but only 11% of its scientists and engineers. Among the minorities, only Asian Americans are represented in the sciences and engineering (4%) beyond their numbers in society (25%; Schmiedler & Michael-Dyer, 1991). Schmiedler and Michael-Dyer commented that

> unless programs are developed to attract and retain more women, minorities, and persons with disabilities into science, mathematics, and engineering (an estimated 85 percent will have to come from these groups), the nation will not be able to meet its technical personnel needs into the next century. (1991, p. 6)

Compounding this problem is the fact that the proportion of women and minorities in the population is increasing; in the year 2020, 24% of the population will be Hispanic, as compared with 9% in 1984. However, only 53% of the population will be White, as compared with 74% in 1984. Thus, American society will become more and more dependent on women and minorities to provide leadership in the sciences.

Mathematically Gifted

The Study of Mathematically Precocious Youth (SMPY), conducted at the Johns Hopkins University by Julian Stanley (1988), studied the demographic characteristics of gifted mathematics students. Stanley and his colleagues have made great strides in understanding the nature of students who perform well on standardized tests through their study of students who score 700 or above on the Scholastic Aptitude Test (SAT) Mathematics subtest at the age of 12 years. In one study of 292 such students, Stanley found the following trends:

- The ratio of male to female students in the group was 12:1 (269 male and 23 female students).
- Seventy percent of the female and 63% of the male students were oldest children; 17% of female and 11% of male students were only children.
- The children's parents were generally well educated.
- The children's verbal skills (also measured by the SAT) were markedly lower than their mathematics skills, but were still superior to those of the average college-bound senior.
- Twenty-two percent of the entire sample were Asian American; only 2% of the students were Black.
- Most of the students also had very high general reasoning ability, as measured by the Raven Progressive Matrices.

Although some of these characteristics, especially those describing gender, race, and family background, may be affected by current cultural barriers to girls, minorities, and economically disadvantaged students, it is useful to have a description of what the current pool of mathematically gifted students "looks like." One additional interesting finding was that the truly exceptional student of mathematics tended to have advanced skills in other areas as well:

> Sheer high IQ alone is not sufficient. SMPY has observed repeatedly that nonverbal reasoning ability of the sort measured by the advanced (36-item) form of the Raven Progressive Matrices also seems important. (Stanley, 1985, p. 207)

This observation raises one of the more interesting quandaries in the education of the gifted—the identification of the mathematically gifted. Often, the identification measures used by school systems involve achievement tests that measure mostly computational skills and not the more complex and sophisticated skills identified by Krutetskii (1976) and mentioned by Stanley. Greenes (1981) pointed out that one result is that sometimes students who are "good exercise doers" but do not have the advanced mathematics reasoning ability qualify for special mathematics programs. If computation-based tests are used to identify students for a program that emphasizes mathematics sophistication, it is possible that these "good exercise doers" will find themselves in a program far beyond their reach and experience damaging disillusionment and failure. Greenes recommended looking for the following kinds of skills when identifying the mathematically gifted: spontaneous formulation of problems, flexibility in handling data, mental agility and fluency of ideas, data organization ability, originality of interpretation, ability to transfer ideas, and ability to generalize.

Equality Versus Excellence and Political Correctness

The national education goals are evidence of a continuing ambivalence in the American public between the contrasting educational goals of *equality* and *excellence*. Three decades ago, John Gardner (1961) raised the question of whether the goals of equality and excellence could be reached simultaneously. He thought it possible, although it was not being accomplished at that time. The passage of time, however, does not seem to have greatly improved the situation. The controversy over equality versus excellence reveals a problem of values. With regard to education, gifted students find themselves in the middle of major value conflicts within American society.

The Fairness of It All

One of the most elusive, but most powerful, inhibitors of programs for gifted students is the value issue of *fairness* and *equality*. Many people seem to wonder whether special programs for gifted students fit into their own value systems. The feeling is often expressed in this way: "Is it really fair for some children to have so much ability and others have so little?

Is it fair for us to be giving special education opportunities to students who already have so many advantages? Is offering these types of special educational opportunities to such students not akin to giving tax breaks to the rich?"

Such problems seem to be made worse by the additional realization that minority groups have a lesser presence in programs for gifted students than their proportion in the general population, with the significant exception of Asian Americans (Zappia, 1989). This lack of minority participation enhances the perception that programs for gifted are really designed as "special privileges for special people" (Baldwin, 1987).

The only answer to all of these value questions is that, "Of course, it is not fair." Abilities are not equally distributed, nor are the opportunities to enhance aptitudes that are present in the child and family. This is not, however, the only unfair thing in the world. It is unfair that so many people live in poverty and disease-ridden environments, whereas others live in opulent wealth. It is unfair that we continue to have wars and that many people are killed needlessly. It is unfair that some countries have continuous droughts, while others prosper under good growing seasons for their crops.

Who among us has the potential to do something constructive to combat this massive unfairness? The gifted students will be among those who have the best chance, when properly educated, to do something about this array of social problems that will face future generations. Just as society supports medical schools and law schools out of enlightened self-interest, because we may need a good doctor or lawyer someday, our enlightened self-interest argues for a solid preparation for the most talented of our students.

The National Educational Goals

This ambivalence, or attempt to achieve the two apparently competing goals of equality and excellence simultaneously, is evident in the national educational goals established by the Governor's Task Force on Education and endorsed by President Bush, and accepted as targets toward which the system should strive for the year 2000. Table 2 indicates these goals. Goals 3 and 4, requiring high competence in content fields and promising

TABLE 2

National Education Goals

1. By the year 2000, all children in America will start school ready to learn.

2. By the year 2000, we will increase the percentage of students graduating from high school to at least 90%.

3. By the year 2000, American students will leave grades four, eight, and twelve having demonstrated competency over challenging subject matter, including English, mathematics, science, history, and geography.

4. By the year 2000, U.S. students will be first in the world in mathematics and science achievement.

5. By the year 2000, every adult American will be literate and possess the knowledge and skills necessary to compete in a global economy and exercise the rights and responsibilities of citizenship.

6. By the year 2000, every school in America will be free of drugs and violence and offer a disciplined environment conducive to learning.

Note. From *America 2000: Our Education Strategy*, 1991, Washington, DC: U.S. Department of Education.

top performance in mathematics and science, represent a major emphasis on excellence and would be highly relevant to gifted students. Goals 1 and 2, in contrast ("that all children will start school ready to learn" and "that 90% of the children will graduate from high school"), represent efforts at achieving equality (U.S. Department of Education, 1991).

There are strong threads in our cultural heritage inclining us toward equality. Many of our ancestors broke away from an elitist society in Europe. Our most treasured documents, the Declaration of Independence and the Constitution, take great pains to ensure that power will not once again reside in the hands of a small elitist group. Many people are loathe to do anything that they believe would strengthen elitist tendencies.

The drive for excellence, in contrast to equality, seems to be based on societal needs. In the modern, postindustrial, information society into which we are emerging, the need for large numbers of well-educated and extensively prepared students is manifest, as is the need for a large pool of creative scientists, managers, communicators, and so forth. The education of gifted students is clearly a high priority for such a society. Unfortunately, the messages that we are now receiving about current student performance are pessimistic, when American students are compared with those of other advanced nations (Crosswhite et al., 1985; Jones, 1988).

Political Correctness and Equality

In reviewing social values that are potentially inhibitory to effective education of high-ability students, there is a strong flavor of political correctness and equality. Political correctness in educational circles means the following, in my view:

- There shall be no differences between sexes, races, and ethnic groups on measures of ability, aptitude, and educational opportunity.
- Any such differences that might appear between such groups are due to flaws in the measuring instruments or in the use of improper or biased instruments.
- Special educational programs should have membership from various subgroups reflecting each group's demographic proportions in the general population.
- Poor performance on the part of members of groups with a prior history of discrimination can be explained by the lack of opportunity that such discrimination represents; this should be compensated for by equitable placement decisions.

Such principles are abandoned only when there is manifest production required, such as in athletics and in music, where such rules could create some distressing setbacks. Think about a U.S. Olympic team, for example, that had to be composed according to the ethnic and racial proportions of society—or a jazz combo or a rock and roll band.

When data do not fit these politically correct values or expectations, there is considerable consternation in some domains of the educational and social science community. Such consternation can lead to lack of support for programs for gifted mathematics and science students unless these demographic proportions are grafted onto the educational program.

In the public schools, leaders of programs for gifted students have been searching, in various degrees of desperation, for alternative selection instruments or procedures that would mitigate the imbalance in the demographic portrait of eligible children in programs for gifted students. The reason for the desperation is the clear, and accurate, perception on the part of such leaders that if the proportions of students depart from the politically correct proportions, then their programs will be attacked as elitist and be under real threat of dismemberment, as indeed they now are (Frasier, 1987).

The idea that some students could be born with more ability in certain intellectual domains than others (Plomin, 1988) or that some children would have what the Russian mathematician Krutetskii (1976) referred to as a "mathematical cast of mind," is disturbing enough to the defenders of equality in all domains, personal and social.

The distressing part of this issue is that developmental psychology has well-documented explanations for this differential performance by race and gender, and that such explanations have nothing to do with any inherent or genetic differences between races or genders. The simple model (see Figure 1) presented by Sameroff (1986) indicates clearly the progressive limitations placed on the development of children by a stimulus-poor environment. If this model is correct, it would be astonishing if the demographic portrait of any applicant pool would be a fair balance because the opportunities for intellectual stimulation are hardly distrib-

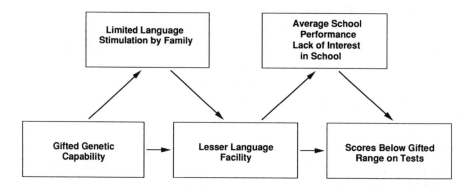

FIGURE 1. Underserved gifted sequence of development. After Sameroff (1986); reprinted with the permission of Cambridge University Press.

uted equally. Such administrative attempts to impose fairness can, and do, inhibit actions that would bring about a true equality.

Educational Adaptations

One of the most common of special educational adaptations for gifted students in mathematics and science is to group them together for a part of the day, or for the entire day, for advanced instruction. The basis for this decision is that (a) students of high ability can learn from one another, (b) the teacher can pace the material at a more rapid rate without worrying about the slower learning children, and (c) a highly qualified mathematics or science teacher can be assigned to this group.

The most common secondary school program is the advanced placement class or honors class, both of which provide an advanced college curriculum, often for college credit. These courses provide the ability grouping and advanced instruction that seem to be useful. At the elementary level, it is still common to use a resource room for one hour a day for bright children despite the limited evidence for its effectiveness (Gallagher, 1985). The teacher of gifted students does not necessarily have any specialized knowledge in mathematics or science. Although such grouping allows for advanced instruction, it also appears to run afoul of current politically correct values, and has met with resistance among leaders of the middle school movement (George, 1988).

A recent meta-analysis on the effects of ability grouping and gifted students was completed by Kulik and Kulik (1991). They summarized their findings as follows:

> The evidence is clear that high aptitude and gifted students benefit academically from programs that provide separate instruction for them. Academic benefits are positive, but small, when the grouping is done as a part of a broader program for students of all abilities. Benefits are positive and moderate in programs that are especially designed for gifted students. Academic benefits are striking and large in programs of acceleration for gifted students. (Kulik & Kulik, 1991, p. 191)

It would appear that merely grouping gifted students together without, at the same time, changing the content and the instructional strategies, will not yield much in the way of benefits. On the other hand, a well-constructed program that brings gifted students together and provides them with an intellectually stimulating and important set of ideas, together with giving them practice to use their own ability to problem find and problem solve, seems to yield tangible results.

Curriculum Adaptation—Study of Mathematically Precocious Youth

One of the most active translators of psychological data to educational purposes in modern times has been Julian Stanley, who some 20 years ago began the Study of Mathematically Precocious Youth at Johns Hopkins University (Stanley, 1985).

Using existing tools, such as the SAT, and knowledge about the results of acceleration of gifted students, which was almost invariably positive (Gallagher, 1985), Stanley devised a program that allowed students with outstanding mathematical aptitude to pursue—through summer programs and through permission to take advanced classes at the college level—a more rapid progress through their educational program.

This program had such attraction that similar centers were established at other universities (Duke, Northwestern, Denver, and Arizona), all of which reported similar broad levels of success in helping students of high ability to pass more rapidly through their mathematics programs and to enter their chosen careers at an earlier age. As was true in the earlier literature, there seem to be few major emotional problems or social difficulties for such students that would counterindicate the program when applied to qualified students.

As one student who had been through the rapid acceleration program put it, "I didn't have much in common with my peers anyway. They weren't interested in the things I was, and now I have a five-year head start on my career." Despite these manifestly favorable results, educators are still hesitant to use acceleration as a strategy for dealing with the advanced cognitive skills of such students (Southern, Jones, & Fiscus, 1989).

Those in favor of acceleration will often quote the literature, which relates highly positive findings (Gallagher, 1985). Those who are against it will respond with a case history of unfortunate outcomes for a given student. Both are, in their way, correct. Still, when used prudently, student acceleration is clearly one strategy for challenging the advanced thinking processes of these students.

Adaptation—Special Schools

Among the newer educational adaptations are the special schools for talented students in mathematics and science that have been established in 10 states, with more on the drawing board (Kolloff, 1991). These residential schools invite statewide applications and the students who are chosen for admission receive an intensive, hands-on, experiment-driven education designed to more adequately approximate good undergraduate and graduate education at the university level. Such residential school programs also serve as a laboratory for the development of alternative curricular design.

The Political Correctness of Educational Reform

Two of the major reform movements in current education are *cooperative learning* and the *middle school movement*. Both carry with them the flavor of equality and democratic processes (all students receive the same opportunities).

The middle school movement has many desirable features to it, blending together much of what is known in developmental psychology. There is an emphasis on the affective life of the student and on personal counseling during this period of "raging hormones," an encouragement of interdisciplinary curriculum (both small-group and large-group work), and an emphasis on deliberate stimulation of thinking processes. Somewhere in the midst of all of these desirable elements, *heterogeneous grouping* has been inserted as a strong middle school component. This blending together of students performing at third-grade level with those at ninth-grade level is a challenge to any teacher or group of teachers. This middle school, heterogeneous grouping philosophy has been used to encourage schools to abandon honors classes in mathematics and science and other areas—hardly the strategy to adopt to meet the year 2000 goals.

Cooperative learning has been sponsored as an instructional strategy by Slavin (1990), Johnson and Johnson (1990), Kagan (1988), and others. Cooperative learning also has many attractive features to it—the experience of learning in a group setting, of team work in investigations, and so forth. The use of cooperative learning has stressed heterogeneous grouping as well, although not exclusively. A small group of four to six students assigned to do a particular task may include one or two of the most advanced students, some average students, and some below average students. The philosophy here is that the more advanced students can help the less advanced students and that overall group performance determines the grade as a motivating device for such help.

As Robinson (1990) has pointed out, there are three substantial drawbacks for gifted students in such arrangements, particularly if cooperative learning lessons are seen as substituting for a particular enrichment program for gifted students:

1. The task chosen for a group activity has to be within the range of the lower students, and often is placed at a basic skills level.
2. The evaluation of performance is usually set at a low level of competence, thus not challenging the gifted students.
3. This low level of challenge prevents the gifted students from stretching their own intellects.

A multitude of parental feedback on cooperative learning as it is being applied in the schools is becoming available, much of it noting individual instances of groups allowing the advanced students to do all the work, complaining that the advanced students are "bossy" as they try to get the job done, or experiencing boredom at the elementary nature of the tasks, and so on.

But both of these strategies, the middle school movement and cooperative learning, fit a distorted concept of political correctness by not allowing any students to progress beyond the level of the overall group, unless, of course, they are engaged in athletics, music, or drama in the school activities program.

Educators, who would consider ridiculous the idea that the best way to educate the next Rubinstein or Horowitz at the piano would be to include the more advanced students with the most elementary ones still learning their scales, or that the ideal training for the next Jack Nicholas would be to have him learning in a group in which the majority of the students are in the woods or lakes of the golf course, are the same educators who would say that gifted students will learn best in heterogeneous settings.

One important bonus for harried educational administrators who want to believe that the heterogeneous grouping idea would work to everyone's advantage is that they would not have to face the embarrassing racial and sexual imbalances that occur in the membership of advanced mathematics and science classes, making them potential targets for discrimination charges.

Underserved Populations

Cultural Differences

Until recently, one of the most embarrassing secrets in the education of gifted students was the differential prevalence of ethnic and racial groups in identification and in placement in special programs. The embarrassment stemmed from the inappropriate assumption that intelligence tests measured only genetic potential and that such a difference in proportions would then suggest superiority or inferiority in the native ability of such groups, thus presenting an intolerable political problem.

Although the objective fact was that there were fewer minority students being identified through traditional methods (except for Asian Americans), the reasons for such low numbers were not universally agreed on. Two major explanations are given for such results:

1. *The instruments and procedures used for identifying gifted students are flawed and biased against those students who are not middle-class, White Americans.* Such an argument rests on the proposition that there should be no true differences in levels of aptitude at the time of assessment, so that any group differences found are the fault of the measurement. Furthermore, the choice of gifted students from the mainstream culture for special educational programs may be viewed as an attempt, possibly deliberate, to limit the opportunities of children from some minority groups. The intelligence tests that have been used by schools may be referred to more aptly as academic aptitude tests and

their predictions of lower performance for minorities as a group have, unfortunately, turned out to be correct for many minority students.

The potential bias of test instruments needs to be demonstrated by more than group differences on the tests. Just as there may be differences between ethnic and racial groups in athletic or musical aptitudes based on greater opportunity and experience, the same may be true of academic aptitude. The excellent performance of Asian Americans on both tests and on school performance indicates that there are factors operating that go beyond simple differences from the mainstream culture (Zappia, 1989).

Nevertheless, the current style of identification tries to cope with this issue by adopting multiple criteria for giftedness, of which IQ tests are only one and not necessarily the determining factor.

2. *This differential prevalence reflects differential opportunities and limited practice on key elements of intellectual development.*

There is considerable evidence to support the importance of the role that practice and experience play in later measures of aptitude (Gallagher & Ramey, 1987). If one can extend the general principle that "we are good at what we practice" to include "we avoid tasks at which we perceive ourselves as not competent, or situations in which we are not comfortable," then it is not hard to see how, progressively, some minority students who may have begun life with equal aptitudes as their majority group age-mates, in terms of a responsive central nervous system, will fall farther and farther behind on measures of academic proficiency and aptitude. Such evidence of differential prevalence, the argument continues, does not address differences in native ability so much as it does differences in the availability of responsive environments to crystallize an individual's native ability.

The most reasonable position, given current knowledge, is to accept Explanation 2—different experiences and opportunities are what make the difference—and operate as though it were true. The obvious step to be taken, then, is early and intensive provision of experiences that can help talented minority students and gifted girls to develop their potential more fully.

Gifted Girls

One of the major groups of underserved gifted students is gifted girls. They are traditionally less represented in programs for gifted and talented, particularly in programs in mathematics and science (Stanley & Benbow, 1986). The traditional roles of women as childbearers and homemakers have clearly been modified, but the new freedom has not yet resulted in remarkable change. Reis and Callahan (1989) pointed out how far society needs to progress:

> Why, for example, are less than 2% of American patentees women? . . . Why are there only two females in the United States Senate, one female on the Supreme Court, and one female cabinet member? Why do women constitute less than 5% of the House of Representatives, own only 7% of all businesses in the country . . . occupy only 5% of executive positions in American corporations, hold none of the leading positions in the top five orchestras in the United States? (pp. 101–102)

Girls with outstanding potential would seem to be the largest untapped resource in our country at this time. One outstanding statistic is that although women make up 51% of the population, they constitute only 11% of the scientists and engineers in the United States, reflecting the vocational and societal tilt against women in these occupations (Schmiedler & Michael-Dyer, 1991).

The idea that a fundamental difference between the sexes would be discovered in some areas of intellectual or academic performance strikes many people as almost un-American. It is *politically incorrect*. Yet, the findings in the area of mathematics are clear: Girls of all ability levels consistently perform less well on measures of mathematics achievement or aptitude than do boys. This is no less true of gifted girls than of girls of average ability. What remains unclear is the reason (or reasons) why.

Research in the area of mathematics and gender has still not provided a definitive answer. Part of the reason is that the question of nature versus nurture is virtually impossible to unravel by the time children are old enough to study mathematics. The two major arguments in the research debate are that (a) the achievement difference is due to some genetic or biological difference, or (b) the difference is due to environmental factors supported by a cultural belief that mathematics is a part of the world of men.

The biological difference side of the issue received its broadest publicity after the publication of an article by Stanley and Benbow (1986). They reported the results of almost 40,000 seventh-grade students from the Middle Atlantic region of the United States who took the SAT as part of the SMPY talent search in 1980, 1981, and 1982. With a relatively equal number of boys and girls sitting for the test, the average scores on mathematics clearly favored the boys by an average of 30 points.

Even more decisive were the results at the highest levels of the test. When the students who scored 700 or higher on the SAT (an extraordinary performance for 13-year-olds) were studied, it was found that 260 boys made this high score, whereas only 20 girls reached that level of performance. This results in a startling 13:1 ratio of boys to girls. Benbow and Stanley (1983) hypothesized that such results may be related to differences in the ability of boys and girls to perceive spatial relationship—that is, to see shapes and patterns.

Although no one can deny the possibility that there may be some biological or genetic differences that account for a part of the infamous "math gap" (although the research shows that the impact of this difference looks smaller and smaller all the time; Linn & Peterson, 1986), it is also hard to deny the cultural hypothesis.

The cultural hypothesis claims basically that the deck is stacked against girls where mathematics is concerned. Mathematics has long had the label of being a masculine activity and not the proper venue for girls. This difference is particularly interesting when one considers that the verbal skills of boys and girls, which once tended to favor girls, have pretty much equalized. Part of the reason for the continued mathematics discrepancy may be differential access to information about mathematics. As Colangelo (1988) pointed out,

> if you don't learn mathematics in a formal mathematics course, you are not going to learn it, because it's not the kind of thing that kids down the street talk about . . . whereas in English, and especially in Social Studies, there

are things you can learn by watching television, by chatting with others, or by reading *Time* magazine. (p. 236)

In a study of more than 100 fourth- to sixth-grade classes, Sadker and Sadker (1985) found that boys still dominate in the classroom and are still encouraged to dominate, whereas girls are encouraged to sit still and behave. In fact, the most surprising of their findings was that of all of the groups studied, high-achieving girls received the least attention.

One natural question that emerges from this difference is whether there are classrooms in which girls are encouraged to perform. Eccles (reported in Reis & Callahan, 1989) did find classrooms in which girls had relatively higher levels of achievement, and she studied the characteristics of those classrooms. What she found was a set of classroom qualities that affected positively girls' confidence and fostered positive attitudes toward mathematics, leading to their equivalent achievement. The characteristics of these classrooms included

- frequent use of cooperative learning opportunities,
- frequent individualized learning opportunities,
- use of practical problems in assignments,
- frequent use of hands-on opportunities,
- active career and educational guidance,
- infrequent use of competitive motivational strategies,
- frequent activities oriented toward broadening views of mathematics in physical sciences, presenting mathematics as a tool for solving problems through frequent use of strategies to ensure full class participation.

Another idea that has proven successful is to group girls together so that boys do not have an opportunity to dominate and the girls cannot rely on the boys for help. In studies of sex-segregated classrooms, it has been found that these girls make greater gains than those in traditional settings (Fox, 1977; Lee & Byrk, 1986).

Studying the lives and accomplishments of female mathematicians also helps provide role models of achieving females. By infusing the works of these women into the curriculum, talented females can learn that they are just as entitled to join the world of mathematicians as are males.

Another area of concern is that of girls and the use of computers. Like mathematics, computers are perceived as a male realm, and girls are less likely than boys to take advantage of the learning opportunities that computers provide. In an interesting book titled *The Neuter Computer*, Sanders and Stone (1986) summarized some of the reasons why men continue to dominate:

- Girls see men operating computers more often than women.
- Computers are related to mathematics, hence, to masculinity.
- Computers are perceived as machinery (not as a tool), hence as masculine.
- Computer software is male oriented, with emphasis on competition, war, death, and sports.
- The assertiveness of boys in the classroom doubles where the computer is concerned—boys leap at the chance to use them, whereas girls patiently wait their turn (which often never comes) or will actually give up time on the computer to boys.

The desire to engage the interest of girls in the sciences has prompted investigations into the motivation and attitudes of gifted students toward scientific careers. Subotnik (1988) investigated the attitudes of 146 winners of the 1983 Westinghouse Science Talent Search, because these would, most likely, make up a group of potential future scientists. These students were given a choice, with regard to their motivation, among curiosity, prestige, aesthetics, and bettering the human condition. Both boys and girls overwhelmingly chose curiosity as their major motivation for doing research, although the girls, significantly more than the boys, chose bettering the human condition as a possible motivation.

Another gender difference was found in the response to the question as to what qualities they most admire in their mathematical or scientific hero or heroine. Responses to this question are presented in Figure 2. Not one of the 20 girls identified creativity as an admired quality in a scientist, with the majority of them responding to a dedication to work as the most admired characteristic. In contrast, intelligence and creativity

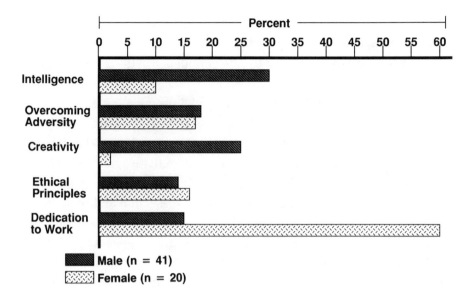

FIGURE 2. Qualities admired in a scientific hero or heroine (Westinghouse Talent Search winners). After Subotnik (1988).

were the two characteristics most admired by the boys in their scientific heroes.

It seems clear that girls entering science have a more humanistic concept and see devotion to work as the key characteristic of a scientist, whereas boys seem more intrigued by the task itself and the opportunity to do creative work. The humanistic interest of the girls in this study is consistent with that in other research, which indicates that girls tend to like to analyze data by making relationships, whereas boys tend to like to analyze by making distinctions (Belenky, Clinchy, Goldberger, & Tarule, 1986). Nevertheless, 57% of these excellent science students reported a desire to investigate relationships between science and society, a topic that receives little attention in school science curriculum.

One common problem that faces even the science-gifted girls is the traditional interactions with gifted boys that they have had in academic programs. Previous experience and observations have revealed that gifted girls often feel uncomfortable with the manipulation of tools and equipment and that the problem is compounded by gifted boys' tendency to commandeer the equipment. The assertiveness of boys and the tacit social understanding that boys are entitled to the use of such equipment does not allow the girls to practice or to gain confidence in the use of materials and equipment.

Rand and Gibb (1989) reported a model program for gifted girls in science that attempted to counteract these issues. Their program is for girls only and has, as a part of the program, a female role model as teacher, family participation, a collection of hands-on activities, and results that give a feeling of success. In this Action Science program, for example, the girls are encouraged to take part in squid dissection and rock collection and classification as two examples of the use of dissection tools and equipment to gather rock specimens. This program appears to have had success in the new-found enthusiasm of many girls for science and in the feeling that they can do many things for themselves in terms of manipulating materials and equipment. As a result of this and other programs focusing on gifted girls, Rand and Gibb (1989) offered four suggestions that they believed would strengthen the interest of these girls in science:

1. *Do not overhelp young girls.* Let them gain valuable experience by thinking through a problem and trying various solutions. Encourage self-reliance or independence in girls.
2. *Encourage girls to trust their own judgment.* Discourage girls from seeking constant approval or verification from others before making decisions or moving to the next step. They need to develop confidence in their own abilities.
3. *Insist that girls use tools and equipment.* Build confidence in their ability to identify and use equipment from basic hand tools to sophisticated computers and microscopes.
4. *Introduce female role models whenever possible.* Include the study of historical as well as of contemporary women in science fields. Remember, also, that females such as mothers and teachers exert a strong influence on a young girl's life.

One final caution needs to be made. Gifted girls can be motivated, reinforced, and programmed until they behave and think as professional scientists do, but only half the problem will be solved. Gifted boys, too, need to be taught to appreciate the contributions of female scientists and of their female classmates. Only when boys and girls alike learn to share roles and responsibilities in the science classroom will the proper environment be achieved to encourage all students to explore their interests and abilities in this fascinating field.

The amount of attention now being received by gifted girls promises more effective programs for them in mathematics and science in the near future.

Future Research Directions

Horowitz and O'Brien (1986) edited a major literature review of gifted students sponsored by the American Psychological Association and concluded with some suggestions for fruitful research.

Understanding Intellectual Processes

This understanding would require investigations of knowledge acquisition, storage, and retrieval, as well as problem identification and solution. Efforts to describe these information-processing mechanisms should extend across the life span.

Differentiating Social and Personality Characteristics

This differentiation would include investigations to determine why some highly intelligent individuals lead concomitantly creative and productive lives, whereas others do not. Variables of socialization, motivation, energy, and personal perceptions appear to influence the degree to which intellectual gifts are fully realized. It is important to look at such characteristics across substantial periods of time.

Assessing Educational Strategies for Gifted Students

We need to determine what kinds of programs most benefit what kinds of gifted and talented children, so that we can better target our scarce educational resources. We should support programs to the degree to which they give evidence that they make a real difference.

Information Processing

One of the most potentially fruitful lines of investigation seems to be to continue and extend the various investigations on information processing in human beings, particularly in children. Little has been written about the "executive function" or the control mechanisms for paying attention, or for choosing among cognitive strategies, or for deciding on a mode of intellectual expression (Borkowski & Kurtz, 1987). Decision making is a poorly understood process, from an information-processing standpoint, and one that could be studied fruitfully in young children, in whom it can be seen in a more observable process than in the complex network of forces affecting decision making in adults.

One element of the executive function operation is the area of problem finding, or of choosing the most significant problem to be attacked. This is important for researchers, but also for politicians, artists, and parents. The right choice can lead to significant findings or products; the wrong choice can lead to months, or even years, of wasted effort. A key area of investigation involves how this process of decision making works and how it can be enhanced.

Family Support

A significant body of investigations exists demonstrating the importance of family encouragement and support for the full development of the intellectual capabilities of talented youngsters. One line of investigation would be how to provide support for families who are not now encouraging their talented youth, in the hope that they would begin to play this role more assertively. Another line of investigation is whether other persons in the environment of the child (friends, relatives, teachers, etc.) can provide the type of support and encouragement necessary to promote full development of these talents if the parents are for some reason unable to do it.

School Program

When evaluating the impact of a particular school environment, such as the resource room (Vaughn, Feldhusen, & Asher, 1991), or ability grouping, or a particular instructional method, such as creative problem solving, one is impressed by the range and diversity of the results. It is clear that resource rooms work well sometimes and not at all well at others. The "enrichment triad" is a great success in some places and a disappointment in others. The mere placing of youngsters in a particular setting, or providing them with a particular set of activities, does not necessarily lead to success. Therefore, it would seem most important to document, in detail, what works.

If a resource room is doing an outstanding job by all accounts, then the particular way in which it is operating needs to be analyzed and studied to understand the ingredients of this recipe for success. If an honors course in philosophy is achieving visible and tangible success, then the nature of that total setting needs to be examined. Is it merely a creative teacher or are there other elements in the situation that need to be recognized? By studying the staffing patterns, the history, the processes, the teachers, and the students, it may be possible to emerge with a better idea of what the recipe for success is within a given structure or program.

Societal Interests

Many of the adaptation problems of gifted students and gifted adults come from the love–hate relationship that such talent generates in the larger society. Socrates, Galileo, and many others have demonstrated what happens to the talented person who runs astray of the larger society or power groups within the society.

It seems reasonable to suspect that attracting envy has always been part of the price that the talented person pays for the expression of his or her talent. With Bach and Verdi, this was probably not terribly inhibiting because they needed to please only relatively few people to continue doing what they wanted to do.

In a democracy, where large numbers of people have a voice in what happens, it becomes increasingly important to understand societal ebbs and flows in attitudes toward these students and adults. What are the dynamics of societal concerns and reservations about such individuals? Is it fear that gifted individuals will use their abilities to gain control over them? How can such feelings be counteracted?

Conclusion

The past 25 years have seen a quantum leap in our understanding of gifted or talented students. Many myths have been dispelled. There has been an increased level of sophistication in the nature of high intelligence itself, as well as in the educational methods that can enhance its development. As more becomes known about giftedness, there has been a greater emphasis on some of the subgroups with special needs, and that emphasis will continue into the near future. One thing that has not changed much has been the ambivalent societal view of how giftedness and gifted individuals should fit into a democratic society.

Gifted behavior may still be viewed as an uncomfortable presence as well as a great advantage. Still, we can increasingly see that we may deny its presence in our youth at our own national peril. We are neither so rich nor so blessed with natural resources that we can, as a nation, afford to ignore educationally the human potential that is embodied in the minds of gifted students.

Our generation will place its signature on the poetry, the science, the art, and the business prosperity of the next generation, in large measure, by how enthusiastically we respond to the educational challenge of these students.

References

Baldwin, A. (1987). Undiscovered diamonds. *Journal for the Education of the Gifted, 10,* 271–286.

Belenky, M. F., Clinchy, B. M., Goldberger, N. R., & Tarule, J. M. (1986). *Women's ways of knowing: The development of self, voice, and mind,* New York: Basic Books.

Benbow, C. P., & Stanley, J. C. (Eds.). (1983). *Academic precocity: Aspects of its development.* Baltimore: Johns Hopkins University Press.

Bloom, B. (1985). *Developing talent in young people.* New York: Ballantine Books.

Borkowski, J., & Kurtz, B. (1987). Metacognition and executive control. In J. Borkowski & J. Day (Eds.), *Cognition in special children: Comparative approaches to retardation, learning disabilities, and giftedness* (pp. 123–152). Norwood, NJ: Ablex.

Bruner, J. (1960). *The process of education.* Cambridge, MA: Harvard University Press.

Colangelo, N. (1988). Discussant reaction: Bright girls in math and engineering. In J. L. Dreyden, S. A. Gallagher, G. E. Stanley, & R. N. Sayer (Eds.), *Developing talent in mathematics, science, and technology: Talent Identification Program/National Sci-*

ence *Foundation Conference on Academic Talent* (p. 236). Washington, DC: National Science Foundation.

Crosswhite, F., Dossey, J., Swafford, J., McKnight, C., Cooney, T., & Travers, W. (1985). *Second International Mathematics Study: Summary report for the United States.* Champaign, IL: Stipes.

Feldman, D. (1986). *Nature's gambit.* New York: Basic Books.

Fox, L. H. (1977). *The effects of sex role socialization on mathematics participation and achievement* (Contract No. FN17400-76-0114). Washington, DC: National Institute of Education.

Frasier, M. (1987). The identification of gifted black students: Developing new perspectives. *Journal for the Education of the Gifted, 10,* 155–190.

Gallagher, J. J. (1985). *Teaching the gifted child* (3rd ed.). Boston, MA: Allyn & Bacon.

Gallagher, J. J., & Ramey, C. T. (Eds.). (1987). *The malleability of children,* Baltimore, MD: Paul H. Brookes.

Gardner, H. (1985). *Frames of mind: The theory of multiple intelligence.* New York: Basic Books.

Gardner, J. W. (1961). *Excellence: Can we be equal and excellent too?* New York: Harper & Row.

George, P. (1988). Tracking and ability grouping: Which way for the middle school? *Middle School Journal, 20,* 21–28.

Goodlad, J. (1964). *School curriculum reform in the United States.* London: Her Majesty's Stationery Office.

Greenes, C. (1981). Identifying the gifted student in mathematics. *Arithmetic Teacher,* 14–17.

Hollingworth, L. S. (1942). *Children above 180 IQ Stanford-Binet: Origin and development.* New York: World Book.

Horowitz, F., & O'Brien, M. (1986). *The gifted and talented: Developmental perspectives.* Washington, DC: American Psychological Association.

International Association for the Evaluation of Educational Achievement. (1988). *Science achievement in 17 countries: A preliminary report.* New York: Teachers College, Columbia University.

Johnson, D., & Johnson, R. (1990). Social skills for successful group work. *Educational Leadership, 1,* 29–32.

Jones, L. (1988). School achievement trends in mathematics and sciences. In E. Rathkopf (Ed.), *Review of research in education* (Vol. 15, pp. 307–341). Washington, DC: American Educational Research Association.

Kagan, S. (1988). *Cooperative learning: Resources for teachers.* Riverside: University of California.

Kolloff, P. (1991). Special residential high schools. In N. Colangelo & G. A. Davis (Eds.), *Handbook of gifted education* (pp. 206–216). Needham Heights, MA: Allyn & Bacon.

Krutetskii, V. (1976). *The psychology of mathematical abilities in school children* (J. Teller, Trans.). Chicago: University of Chicago Press.

Kulik, J. A., & Kulik, C. C. (1991). Ability grouping and gifted students. In N. Colangelo & G. A. Davis (Eds.), *Handbook of gifted education* (pp. 178–196). Needham Heights, MA: Allyn & Bacon.

Lee, V., & Byrk, A. (1986). Effects of single sex secondary school on student achievement and attitudes. *Journal of Educational Psychology, 78*, 381–395.

Linn, M., & Peterson, A. (1986). A meta-analysis of gender differences in spatial ability: Implications for mathematics and science achievement. In J. Hyde & M. Linn (Eds.), *The psychology of gender: Advances through meta-analyses*. Baltimore, MD: Johns Hopkins University Press.

Massey, W. E. (1989, September). Science education in the United States: What the scientific community can do. *Association Affairs, 91*, 5–921.

Mullis, I., & Jenkins, L. (1988). *The science report card.* Princeton, NJ: Educational Testing Service.

Olszewski, P., Kulieke, M., & Buescher, T. (1987). The influence of the family environment on the development of talent: A literature review. *Journal for the Education of the Gifted, 11*, 6–28.

Plomin, R. (1988). Environment and genes: Determinants of behavior. *American Psychologist, 44*, 105–111.

Rabinowitz, M., & Glaser, R. (1985). Cognitive structure and process in highly competent performance. In F. Horowitz & M. O'Brien (Eds.), *The gifted and talented: Developmental perspectives* (pp. 75–98). Washington, DC: American Psychological Association.

Ramos-Ford, Y., & Gardner, H. (1991). Giftedness from a multiple intelligences perspective. In N. Colangelo & G. Davis (Eds.), *Handbook of gifted education* (pp. 55–64). Needham Heights, MA: Allyn & Bacon.

Rand, D., & Gibb, L. (1989). A model program for gifted girls in science. *Journal for the Education of the Gifted, 12*, 142–155.

Reis, S., & Callahan, C. (1989). Gifted females: They've come a long way—Or have they? *Journal for the Education of the Gifted, 12*, 99–117.

Robinson, A. (1990). Cooperation or exploitation? The argument against cooperative learning for talented students. *Journal for the Education of the Gifted, 14*, 9–27, 31–36.

Sadker, D., & Sadker, M. (1985, March). Sexism in the schoolroom of the 80s. *Psychology Today*, pp. 54–57.

Sameroff, A. (1986). Neo-environmental perspectives on developmental theory. In R. Hodapp, J. Burock, & E. Zigler (Eds.), *Issues in the developmental approach to mental retardation* (pp. 93–130). New York: Cambridge University Press.

Sanders, J., & Stone, A. (1986). *The neuter computer.* New York: Neal-Schuman.

Schmiedler, A., & Michael-Dyer, G. (1991). *State of the scene of science education in the nation.* Washington, DC: Public Health Service.

Simon, H. (1980). Problem solving and education. In D. Tuma & R. Reif, *Problem solving and education: Issues in teaching and research*. Hillsdale, NJ: Erlbaum.

Slavin, R. (1990). Ability grouping, cooperative learning and the gifted. *Journal for the Education of the Gifted, 14*, 3–8, 28–30.

Southern, W., Jones, E., & Fiscus, E. (1989). Practitioner objections to the academic acceleration of gifted children. *Gifted Child Quarterly, 33*, 29–35.

Stanley, J. C. (1985). A baker's dozen of years applying all four aspects of the Study of Mathematically Precocious Youth (SMPY). *Roeper Review, 7*, 172–175.

Stanley, J. C., & Benbow, C. P. (1983). SMPY's first decade: Ten years of posing problems and solving them. *Journal of Special Education, 17*, 11–25.

Stanley, J. C., & Benbow, C. P. (1986). Extremely young college graduates: Evidence of their success. *College and University, 52*, 361–371.

Sternberg, R., & Lubart, T. (1991). Our investment theory of creativity and its development. *Human Development, 34*, 1–31.

Stevenson, H. W., Stigler, J. W., Lee, S. Y., Lucker, G. W., Kitamura, S., & Hsu, C. C. (1985). Cognitive performance and academic achievement of Japanese, Chinese and American children. *Child Development, 56*, 718–734.

Subotnik, R. (1988). The motivation of experiment: A study of gifted adolescents' attitudes toward scientific research. *Journal for the Education of the Gifted, 11*, 19–35.

Terman, L., & Oden, M. (1947). *The gifted child grows up: Twenty-five years of a superior group* (Vol. 4). Stanford, CA: Stanford University Press.

U.S. Department of Education. (1991). *America 2000: An education strategy*. Washington, DC: Author.

Vaughn, V. L., Feldhusen, J. F., & Asher, J. W. (1991). Meta-analyses and review of research on pull-out programs in gifted education. *Gifted Child Quarterly, 35*, 92–98.

Whitehead, A. N. (1929). *The aims of education*. New York: Macmillan.

Zappia, I. (1989). Identification of gifted Hispanic students. In C. J. Maker & S. Schiever (Eds.), *Critical issues in gifted education* (pp.19–26). Austin, TX: Pro-Ed.

PART TWO

Cognitive, Affective, and Situational Factors in Mathematics and Science Education

Roles of Cognition, Emotion, and Social Interaction in Mathematics and Science Education

Douglas L. Nelson

E ducation in the United States is perceived to be in crisis by leaders in government, business, and education. Whereas all agree that education is the key to economic strength and to quality of life, it has become increasingly clear that many citizens are not being trained adequately for the technological demands of today's work force and certainly not for tomorrow's demands. As with any crisis, however, the first step in solving it is to recognize the problem. This crisis was officially recognized at the historic education summit held in Charlottesville, Virginia, in 1989, where the nation's governors met to establish national performance goals for education that would make the country internationally competitive. The second step in any crisis is to apply current resources and knowledge to solve it. As often happens, we already know a great deal as a result of the ongoing research activities of basic and applied researchers. The purpose of this portion of the book is to present the ideas and research findings of several distinguished investigators who have been concerned with applying a portion of what we have learned about cognition, emotion, and social interaction to the classroom. These chapters by no means represent an exhaustive survey of what is known or what can be applied, but they serve to illustrate that research on memory, emotion, and problem solving has direct implications for understanding and improving skill learning in the classroom.

Most language, mathematics, and science skills are acquired in the primary and secondary school years, with little formal rehearsal of those skills in later years. Volumes of laboratory research on forgetting show that unrehearsed information quickly becomes inaccessible. Insofar as conclusions based on this research generalize across all types of knowledge, much of what is learned in the classroom would become useless in a few months or years. Laboratory work on forgetting, however, has focused narrowly on information losses occurring after superficial study, and, although this work is extremely important for settling many theoretical and practical issues about memory, it may not tell us much about what kinds of information are maintained over a lifetime and the conditions that ensure a high level of performance. We obviously remember simple arithmetic skills and the meanings of thousands of words, but this ability may arise because this information is frequently used and, consequently, rehearsed; but what about information that is rarely used? Does one retain any information learned in school over a lifetime? If so, what kinds of information can be retained and what kinds of information are likely to be lost, and what is it about certain types of information and certain training methods that produce retention over very long periods?

These questions form the focus of chapter 2 by H. P. Bahrick. His work shows that some aspects of knowledge become so well learned that access is maintained over a lifetime even in the apparent absence of rehearsal. His approach involves the use of methods long used to study individual differences in achievement, but not to study memory. These methods require the construction of tests for evaluating the recognition and recall of information acquired during high school. For example, he has studied memory for the names and faces of classmates, for Spanish vocabulary and grammar, and for mathematics. Thousands of subjects of different ages were tested and data were obtained from each participant about how well the information was learned originally, about the type and amount of rehearsal, and about the length of the retention interval.

Multiple regression techniques were then used to predict retention scores. These techniques show that unrehearsed information can be recognized for many years after original learning. Even after half a century, people recognized reliably the names and faces of classmates, the meanings of some Spanish nouns and verbs, and certain facts about algebra and geometry. This information appears to be stored permanently. The most important variable in determining the level of performance was how well the information was learned in the first place. Differences in the retention functions for Spanish and mathematics were attributed to differences in the amount and distribution of practice related to the instructional methods used in teaching the two types of material. For example, in contrast to the gradual declines obtained for Spanish, students with college training at or above the calculus level showed little or no loss of mathematics over their lifetimes. This stability is attributed to reviews of the relevant material in each successive mathematics course. Overlearning information by distributing practice over long intervals can result in skills that show only modest declines over a lifetime even in the absence of practice during the retention interval.

Bahrick uses these and other findings to make three important recommendations. First, educators need to begin studies in which retention for course content is measured over very long intervals. The methodology for doing such studies is now available and the results of such studies can be used to evaluate various forms of instruction. Second, strategies for increasing retention over long intervals need to be incorporated into the curriculum. Retention is likely to be increased by using mandatory reviews in subsequent courses and by spreading instruction over longer periods of time. Third, he proposes a new method that relies on repeated testing for maintaining rapid access to newly learned information, called the *method of asymmetry of fluctuations*. The method derives from comparisons between tests of recognition and recall. Although memory for large amounts of information can be demonstrated on tests of recognition, recall appears to be more variable and unstable. Information recalled on one test occasion often cannot be recalled on the next. The asymmetry method is based on findings showing that upward fluctuations on repeated tests of recall generally exceed downward fluctuations and that a single successful recall has a half life of about a month. By using this method, optimal schedules of testing and rehearsal can be determined for maintaining rapid access to marginally learned content. Furthermore, he suggests that repeated testing can transform marginally learned content into permanently stored memories.

Bahrick's findings and his recommendations do not consider the emotional state of the student as learner or as test taker. Like most of the research in the memory field, individual differences in emotional state are left to vary at random on the assumption that they will have little effect on the general principles emerging from the findings. This view is not unreasonable given the need to focus research projects on specific and manageable goals, but we know that students come to the classroom with a variety of attitudes about themselves, their teachers, and the information that is to be mastered. These attitudes are often reflected in emotional reactions that can facilitate or interfere substantially with their progress in acquiring new information or a new skill.

Emotional reactions may influence performance in every learning and testing situation. There may be no such thing as an emotionally neutral state when a person is engaged in learning or in retrieving information. Emotional reactions may be much more prevalent than is presumed. What we do know is that emotional state is correlated with cognitive performance. H. C. Ellis and his colleagues (chapter 3) review six theories on the relationship between cognition and emotion, discuss the implications of these theories for classroom learning, and outline coping strategies designed to alleviate the effects of stress.

The six theories focus on the effects of anxiety and depression on performance and therefore emphasize the negative consequences of the cognitive–emotive relationship. Individuals who are anxious or depressed tend to perform at lower levels on moderately complicated tasks compared with individuals who are less anxious or depressed, and this relationship is captured by various theories in different ways. For example, some theories attribute this performance decrement to reduced attention, increased arousal, decreased processing efficiency, or reduced resource allocation. Other theories emphasize the conceptual aspects of emotionality and treat emotional states as units of memory that can activate other connected units or as conceptual structures for processing and organizing experience. Each theory tends to be concerned with its own domain of facts, but they often overlap and complement rather than contradict one another.

Ellis et al. note that some of the theories have their roots in the study of anxiety and depression, whereas others trace their roots to work on attention and cognition. A combination of approaches appears to work best. For example, resource allocation theory borrows the idea that encoding requires processing capacity from the research literature on attention, where the notion that we have limited capacities to process information was developed. By combining the limited capacity idea with the assumption that emotional states reduce this capacity, the theory explains how the emotional aspects of anxiety and depression can reduce performance on moderately complicated tasks. Furthermore, the theory can explain how worries about performance can intrude and disrupt a task by borrowing the idea that task-irrelevant thoughts increase with increased negative moods from the research on anxiety. Highly anxious individuals often think more about how they are doing than about what they are doing. Resource allocation theory implies that high levels of anxiety and depression can reduce performance because the emotional state directly reduces processing capacity, or because the emotional state fosters intrusive worries, or both.

The review of the literature reported in chapter 3 suggests that classroom performance is influenced by both the emotionality component and the worry component but that, at the present time, the relative influence of each factor cannot be determined. This issue may be important because, as Ellis et al. note, treatment techniques for reducing anxiety tend to focus on one component or the other. For example, systematic desensitization and cognitive therapy tend to focus on the worry component. Cognitive treatments teach students to replace negative thoughts (e.g., Barbie's infamous "Math class is tough") with positive thoughts (e.g., "Math class is fun"). In contrast, relaxation training tends to be directed at the emotionality component alone. Although the findings are preliminary, they indicate that desensitization and cognitive procedures are more effective than relaxation procedures in promoting improved performance.

The research reviewed by Bahrick and Ellis indicates that the ability to learn and remember information experienced in the classroom is a function of the training and testing conditions used and of the emotional state of the individual. In recent years, researchers have come to recognize that social factors play an important role in cognition as well. Learning, retrieving, and thinking are affected by the knowledge and skills that an individual brings to a task at hand, but these activities often take place in an interactive context involving other individuals. In these situations, the cognitive activities of an individual reflect an interaction between the individual and a situation, and these activities may not be uniform across persons or across situations. Individuals and social groups can show fundamental differences in their approaches to learning and thinking that arise as a function of information acquired through social interaction and everyday cultural experience. The chapters presented by John Bransford and his associates in the Cognition and Technology Group at Vanderbilt University and by James Greeno at Stanford University recognize the importance of the social context in knowledge acquisition and stress the idea that students should be active participants in the construction of their own knowledge.

The Cognition and Technology Group at Vanderbilt has been developing and testing educational programs designed to motivate students to learn to think for themselves. These programs incorporate well-established psychological principles, and they have been devised to teach students to solve complex problems that mimic real world events. In their Jasper series, students watch interesting and complicated short stories involving real people and realistic events. At the end of the story, the protagonist is faced with a problem (e.g., a time–distance problem) that must be solved by the students before they see how the characters solve it. Students may then be given a second story that can share many or only a few features with the first, and this manipulation allows these researchers to investigate transfer of training. It is important to know what aspects of the training experience will help or hinder problem solving on related problems. Transfer would be complete if the learning experience enabled students to solve all such problems under all the conditions that they might encounter in the future (e.g., recognizing and solving all time–distance problems regardless of context). This level of transfer would be ideal, but may never be achieved in practice. More likely, transfer will tend to be specific, applying primarily to problems that are similar to the training problems. The specificity of transfer underscores the importance of training students to solve problems in real world contexts as opposed to more traditional word problem methods that present information in sterile contexts often perceived as unrelated to the real world. Such training is not as likely to motivate students, nor is it likely to produce students who can generalize their skills to realistic problems.

Jasper problems overcome the limitations of more traditional methods because they rely on "anchored instruction" and simulate concepts in meaningful contexts. The simulation anchors the problem in memorable, realistic situations. These problems make extensive use of video technology and emphasize complex and open-ended problem solving. Because students work in small groups, the problems encourage communication among students and allow them to generate the solutions themselves. This communication is regarded as a critical aspect of the program. The Vanderbilt group assumes that students cannot learn to engage in effective problem solving when taught by traditional methods that encourage student passivity. The group takes a strong *constructivist* position and argues that students must be actively involved in the construction of their own knowledge. They note that encouraging students to generate their knowledge gives them some of the same advantages that experts have when they are trying to acquire new information that is relevant to their expertise. Experts experience changes in their own thinking when they encounter new ideas. In contrast, novices often experience new ideas as new information that must be memorized because someone says so. By allowing students to generate their own knowledge under the guidance of the teacher and of the program, students can experience changes in their thinking directly, just as the experts do. The Vanderbilt group reasons that generation, as opposed to reception, will produce a student who is more capable of solving complex real world problems encountered after leaving school. Research findings concerning the "generation effect" in the memory literature and the Vanderbilt group's own "baseline" studies provide clear evidence for this claim. Furthermore, because the program has built-in evaluation features, the findings are being used to modify the materials as research findings suggest limitations. For example, by incorporating "what if" variations into the Jasper series as part of the training component, students show higher levels of transfer to new problems.

J. G. Greeno also takes a strong constructivist position on education. He views cognition and learning as arising from an interaction between the learners and other people and physical systems. This view stands in marked contrast to traditional views that treat learning and cognition as individual achievements with social conditions serving merely as part of the context. Instead, Greeno sees individual achievement as being intimately related to group interaction processes. Patterns of interaction that can be learned are called *practices*, and as practices differ from one community to the next, so do individual achievements. This view suggests that the ways in which individuals formulate concepts, construct arguments, use evidence, and create solutions are determined by social interaction. In the constructivist view, learning and cognitive processes are best understood as social activities for increasing participation in community practices. Knowledge of mathematics and science is constructed in discourse between the members of a group, and their progress in understanding can be inferred from conversations among the participants. In the first section of his chapter, Greeno suggests that studies of "communities of practice" associated with commercial transactions, science laboratories, and school functions provide support for the importance of this view. Furthermore, by focusing on the study of practices in realistic situations, researchers will come to understand how cognitive processes are affected by social interactions, and this understanding will lead to intelligent changes in the school curriculum. Such study is also likely to influence basic research on cognition and learning, which has generally ignored the social aspects of mental processing.

Greeno argues that more needs to be learned about how students informally understand science and mathematics before curricular changes in learning activities are designed. Recent work indicates that children have a deeper implicit understanding of fundamental concepts than previously believed. They have informal, implicit theories that become more differentiated and sophisticated as children become more experienced. With experience, implicit understanding involving the use of a concept moves to a level of explicit understanding involving its explanation in language. Understanding progresses from implicit to explicit as it moves from using a concept to explaining it, and this change is facilitated by active social interaction. In implementing this approach, Greeno concludes that a theory of implicit understanding is a necessary step in developing a comprehensive theory of educational practices, and a major portion of his chapter is devoted to describing his work on this task.

His work focuses on children's implicit understanding of the concepts of variable and linear function. They were asked to solve problems with a "winch system" that consists of a short board with two tracks that are marked with distance in inches. Each track has a block that is attached by a string to a spool that can be turned by a handle. Turning the handle moves the block, and different types of problems can be introduced to the students by varying the size of the spool and connectedness of the spools on the two tracks. Students always work in pairs to solve the problems and can communicate freely. By determining how the children arrived at correct solutions to questions posed by the experimenter, their level of implicit understanding of variable and linear function can be inferred. This work shows that some problems were solved as a result of the constraints of the winch system itself, whereas other problems were solved by using arithmetic constraints as well. Interestingly, different child pairings solved arithmetic constraints by different means, with some relying on mental arithmetic and others relying on constructing a table of numbers, pointing and counting, or turning the handle and counting.

This research shows that children have a high level of implicit understanding of variable and linear function, and it suggests two key ideas for constructing a theory. First, situations contain regularities or *constraints*, and they contain environmental supports for various activities or *affordances*. Children are sensitive to the constraints and affordances in many situations and use these characteristics to solve problems. Second, concepts arise in conversations between children to solve problems. These concepts are instrumental to activities involved in planning and goal setting, they can be used as a means for discourse by the children, and they can be used by the community for purposes of determining what the children understand.

Greeno's theory of implicit understanding is a new theory of concepts in which concepts are born of conversations among members of a community responding to constraints and affordances in their environment to solve problems of common interest. The theory goes well beyond theories of concepts in the current literature and can be used to guide curriculum development, to study conceptual growth, and to invigorate the scientific study of concept formation. Greeno suggests that researchers now possess the means to develop models for most routine instructional tasks that will deepen the synthesis of process and content. He stresses the need to place greater emphasis on this synthesis and suggests that this goal can be achieved by combining the study of social interaction with the study of conceptual structures involved in reasoning and problem solving. Educational practice at all levels is likely to benefit substantially by encouraging student inquiry, because conversations help students perceive the constraints and affordances existing in new and challenging tasks. Finally, Greeno notes that educational practice has been profoundly influenced by ideas developed through psychological research. These chapters provide justification for this influence.

CHAPTER 2

Extending the Life Span of Knowledge

Harry P. Bahrick

The relevance of psychological research to education was recognized more than 100 years ago (Herbart, 1824/1982). During the past century, a succession of distinguished scholars of learning have made impressive contributions to education that have led to curricular reform, measurement of and adjustment to individual differences, programming of content, improvements in the writing of textbooks, and changes in pedagogy.

In contrast to the many contributions to education learning research, it is difficult to point to significant contributions of memory scholarship to education. The failure of memory scholars to address the concerns of educators was criticized by Neisser (1978), who described this neglect as scandalous. The lack of impact of memory research on education appears

Preparation of this chapter was supported by National Science Foundation Grant BNS-8417788. The author wishes to thank Lynda Hall and Thomas Nelson for many valuable suggestions, and Lester Krueger and Moshe Naveh-Benjamin for providing data pertaining to the retention of knowledge acquired in a statistics course.

to be paradoxical if one considers that memory research focuses on retention of information and that the value of education depends crucially on the retention of information. How could education fail to benefit from 100 years of memory scholarship? Kintsch (1974) explained this paradox:

> Most of the experimental research concerning memory has never really dealt with problems of the acquisition and retention of knowledge, but with episodic memory which is not at all the problem of interest in education. ...An educational technology squarely based upon psychological research needs research concerned with problems of knowledge. (p. 4)

Memory scholars have focused on episodic content, neglecting semantic content, because semantic content does not lend itself to the dominant experimental paradigm. The experimental paradigm, pioneered by Ebbinghaus (1885/1964), is not suited to covering the long time periods during which semantic knowledge systems are learned and retained. Memory experiments require control of the conditions of acquisition and control of rehearsals during the retention interval. These controls are manageable in laboratory sessions lasting for a few hours, but not under circumstances in which acquisition and retention may extend over months and years. It is not surprising, then, that only a few of the several thousand longitudinal investigations of memory have dealt with content acquired in classrooms.

Naturalistic investigations rely on correlational methods to deal with variables that cannot be manipulated or controlled. Correlational methods easily accommodate long acquisition and retention periods and are therefore well suited for investigating long-term retention of semantic memory content. During the past 20 years, we have used correlational methods to explore the effects of a variety of predictor variables on the life span of knowledge in various domains (including knowledge of high school mathematics and the Spanish language acquired in school). The purpose of this chapter is to describe these methods and the findings that they yield, to discuss their implications for education, and to outline programmatic follow-up research that will allow educators to extend the life span of knowledge at an acceptable cost.

Recent criticism of the naturalistic approach to memory research (Banaji & Crowder, 1989) has led to a discussion of research strategies designed to advance the field. It is not within the purview of this chapter to recapitulate these discussions. It is my view (Bahrick, 1989) that the experimental and naturalistic paradigms are supplementary:

> We must continue to build on the legacy of 100 years of experience, and the evolution of laboratory methodology is the most significant portion of that legacy. New methods will supplement rather than replace the methods now available, and laboratory manipulation will remain preferable to naturalistic observations whenever the phenomena of interest permit it. (p. 82)

This view is shared by other memory scholars (Bruce, 1991; Ceci & Bronfenbrenner, 1991; Conway, 1991; Klatzky, 1991; Loftus, 1991; Neisser, 1991), and it has prevailed in other areas of psychological research (e.g., animal behavior, social psychology) in which both naturalistic and experimental methods are now recognized as supplementary research strategies vital to the advancement of the field. Even in those domains in which naturalistic investigations must be the primary approach, the laboratory will continue to be the place for resolving questions left unanswered by naturalistic observations.

Evidence for Permastore

Correlational investigations of very long-term memory use cross-sectional methodology. A large number of participants are tested for retention of a given memory content; some of the participants acquired the content recently and others as long as 50 years ago. Additional information is obtained from each participant with regard to the level of original knowledge, the amount and type of rehearsals during the retention interval, and the length of the retention interval. A multiple regression analysis is then performed to predict retention scores from indicants of the degree of original knowledge, the amount of rehearsal, and the length of the retention interval. Retention functions are projected by evaluating the regression equation for successive increments of the retention interval.

At the outset, there were serious concerns about the accuracy of subjects' reports of their degree of original learning and rehearsals. These reports include the number and the level of relevant courses taken and grades received, and serve as predictors of retention. Yet, the reports themselves reflect potentially unreliable memory content. Although the control of errors of predictor variables continues to be an important concern, the results show that these errors do not seriously impair either the method or the findings.

The most reassuring evidence concerning the validity of the method is the high multiple correlation achieved in predicting retention performance. A multiple correlation above .80 was attained in predicting retention of high school algebra content over a 50-year period (Bahrick & Hall, 1991a). This multiple correlation is based on a very large number of participants and is therefore stable. The high multiple correlation indicates that the predictors account for a large portion of the variance of retention scores in spite of all potential sources of error. We also verified the number of courses taken and grades received for a large number of participants by comparing the reports with school records, and although participants frequently give somewhat inaccurate reports, the overall validity of the predictors remains high.

Figures 1 and 2 show projected retention functions over a 50-year period for the content of high school and college courses in the Spanish language and for the content of high school algebra, respectively. Both figures also show the effects on retention of the number of relevant courses taken. Projected retention functions are obtained by evaluating the regression equation yielded by the regression analysis. The equation is evaluated for successive increments of the retention interval and for each desired value of the number of courses within the parameters represented by the subjects of the investigation. Thus, the projected functions involve no extrapolations.

Figure 1 shows that retention of the Spanish language declines for a period of 3–5 years, then stabilizes for 25 years before showing further decline during the fifth and sixth decades of life. It is important to note that the projected functions control the effects of rehearsals of content during this long period; that is, rehearsal variables are set at a value of zero in the regression equation used to generate the retention functions.

The original publication on the retention of Spanish language (Bahrick, 1984) also showed traditional retention functions based on the mean

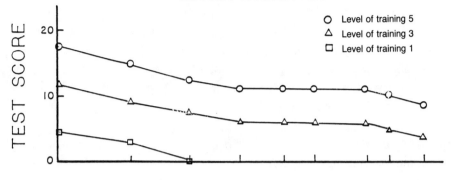

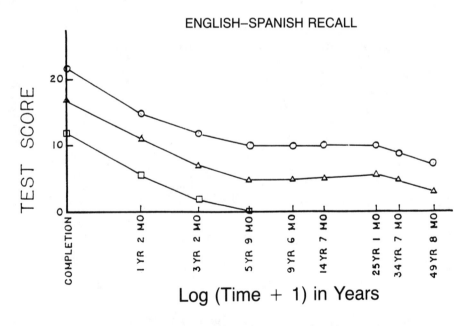

Log (Time + 1) in Years

FIGURE 1. Effect of level of training on the retention of recall vocabulary. Reproduced from Bahrick (1984) by permission.

performance of subgroups of participants tested at various time intervals following their original training. When these functions are adjusted by multiple regression so that the mean retention scores of all subgroups reflect the same degree of original training and zero rehearsal, the functions also show the long period of stable retention seen in Figure 1. However, the performance level at which the function stabilizes becomes an artifact of the mean level of original training (i.e., the average number of courses participants happen to take). This artifact obscures the important influence of the degree of original training on the level of performance stabilization evident in Figure 1. To wit, retention of Spanish–English recall vocabulary stabilizes at a level of 0%, 29%, and 63% of the original score for those who took one, three, or five courses, respectively. Although these percentages are certain to be affected by the difficulty level of the test, the findings underline the strong influence of the level of original training on the portion of content called *permastore content*, (i.e., content that is maintained for several decades without the benefit of additional rehearsal).

Figure 2 shows that the retention of high school algebra content is also strongly affected by the number of subsequent mathematics courses

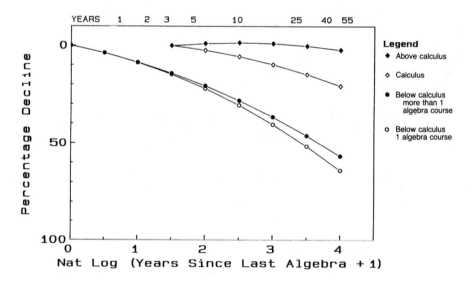

FIGURE 2. Predicted percentage decline in performance on the alegbra 1 test as a function of number and highest level of mathematics courses taken. Reproduced from Bahrick and Hall (1991) by permission.

taken, but stabilization of knowledge is achieved only by those participants who take additional mathematics courses above the level of calculus. The retention functions for Spanish and for algebra differ in several important ways. Retention for Spanish shows a decline during the first 3–5 years of the retention interval regardless of the level of original training of the participants. This decline occurs because some new material (e.g., vocabulary) covered in the test is acquired in each additional Spanish course, and this material does not receive the extended practice required for permastore stability. In contrast, advanced courses in mathematics do not add new high school algebra content. They require review and relearning of the high school algebra content, but the new, nonalgebra material covered is *not* included in the retention test. Individuals with permastore retention of high school algebra, therefore, do not show retention losses during the first 3–5 years of the retention interval. Similarly, the retention function for recognizing names and faces of high school classmates (Bahrick, Bahrick, & Wittlinger, 1975) shows no losses during the early years of the retention interval. Exposure to the material extends over 4 years, and generally few, if any, new names and faces are added during the later portions of the acquisition period.

Permastore stability of the content of high school algebra is achieved only by those participants who use or relearn high school algebra over an extended period in follow-up mathematics courses. Participants who are not reexposed to the content in additional courses show substantial forgetting regardless of their aptitude for mathematics (SAT scores) or of the grade earned in algebra. The difference in permastore retention of Spanish versus algebra is best explained on the basis of the amount and the distribution of practice available in the basic courses for the two disciplines. Beginning language courses provide continuing exposure to basic vocabulary and syntax throughout the course. Thus, practice of certain portions of content is extended throughout the course, and these portions may acquire permastore stability on the basis of a single course in the discipline. In contrast, it takes several mathematics courses in which the high school algebra content is relearned and used before that content stabilizes. Apparently, the amount and distribution of practice obtained in a single course is not as high for the high school algebra content as it is for portions of the content of beginning language courses. However, once the critical level of overlearning is reached in subsequent courses, the content of high school algebra also stabilizes.

Naturalistic investigations do not offer the degree of control of experiments. The greater danger of confounding important variables makes it particularly important to consider alternative explanations of findings. For example, individual-differences variables related to motivation or aptitude were not controlled in the study of retention of algebra content. If those who select additional mathematics courses are also more highly motivated or more talented, or both, these variables could affect permastore retention, independently of the effect of the reexposure to content that additional courses offer. However, the regression analysis shows that neither grades in courses nor aptitude, as reflected by standardized test scores (e.g., SAT), interact significantly with the observed rate of forgetting, and this finding does not support such an interpretation.

It is also possible that acquisition of the content of higher mathematics courses affects retention of high school algebra in ways that go beyond the effect attributable to relearning and to using the basic content. If such additional effects occur, they cannot be sorted out on the basis of the available data.

Both investigations, however, confirm the important conclusion that academic instruction can yield performance that is stable, over many years, on tests of retention for unrehearsed knowledge. This finding is also confirmed by Conway, Cohen, and Stanhope (1991), who reported no decline after the third year of the retention interval in tests of course content in cognitive psychology. It should be noted that the findings of stabilized retention are not limited to a particular type of content or to a particular conceptualization of learning or retention (e.g., associative, reproductive vs. organizational, reconstructive). The retention tests for mathematics and for Spanish language included items that require verbatim retention of specific facts (e.g., vocabulary or formulas) as well as items that require the application of general principles or schemata to content not previously encountered (e.g., testing reading comprehension of Spanish paragraphs or solution of new mathematics problems).

Criticisms of the Permastore Concept

Hintzman (1993) asserted that the evidence for permastore memory is an artifact of grade inflation. His claim was made in a book chapter without the benefit of any supporting data. However, in a widely distributed technical report cited by him as a manuscript in preparation, Hintzman supports his claim on the basis of statistical adjustments to the memory functions reported by Bahrick (1984). He presented evidence of grade inflation in various U.S. universities during the 1960s and 1970s and inferred, on this basis, that students who received their Spanish language training during the early years of the retention interval had more knowledge of Spanish than those who earned the same grade, but were trained later in the interval. By statistically adjusting retention performance to discount the inferred effects of grade inflation, he generated new retention functions that decline during the retention interval.

His principal adjustment was inappropriate because the grade inflation data that he used were not relevant. The Spanish language grades used as predictors by Bahrick were not subject to grade inflation during the critical interval. Two other minor adjustments by Hintzman were also inappropriate because he merged data from various Spanish language tests that were logically, procedurally, and statistically independent, and because he used an inapplicable adjustment for reconciling college and high school Spanish language grades. The stabilized memory functions reported by Conway et al. (1991) gave additional conclusive evidence against Hintzman's assertions (Bahrick, 1992). No high school grades were involved in the Conway et al. investigations, and grades had a trivial effect on memory functions. Therefore, grade inflation is not responsible for the stable retention functions reported, nor is it responsible for the findings of stabilized semantic memory content reported in other investigations (Bahrick et al., 1975; Bahrick & Hall, 1991a).

In contrast to Hintzman (1993), who alleged that stability of unrehearsed content is artifactual and that the laboratory evidence accumulated during the past 100 years indicates that memory functions decline continuously, Banaji and Crowder (1991) asserted that the observed long periods of stability for the retention of academic content tell us nothing new about memory: "One lesson that Bahrick cites is that forgetting proceeds rapidly at first, and then slows down greatly to reach a nonzero asymptote. Most of us thought we already knew the general form of measured forgetting from more than a century of solid research" (p. 78).

Clearly, Hintzman (1993) and Banaji and Crowder (1991) did not draw the same conclusions from the available laboratory findings. Hintzman believed that stability of memory functions is artifactual and inconsistent with the findings of continuous forgetting reported by Ebbinghaus (1885/1964) and others, whereas Banaji and Crowder (1991) claimed that Ebbinghaus (1885/1964) and others demonstrated that memory functions reach nonzero asymptotes.

The contradictory conclusions of Hintzman (1993) and Banaji and Crowder (1991), drawn from 100 years of laboratory findings, illustrate the difficulty of interpreting laboratory findings with regard to the long-term stabilization of retention functions. The functions obtained in the laboratory are based on episodic content, and they cover relatively short time periods. More important, although the laboratory functions are negatively accelerated, they exhibit no period of stability. Inferences of extended stability or instability of content on the basis of mathematical extrapolations are speculative.

What matters to educators, however, is not the question of whether plateaus extending over many years in the retention of unrehearsed academic content could or could not be predicted from laboratory memory work. Rather, the critical concerns of educators pertain to identifying conditions of acquisition that lead to stabilized retention functions and to determining the duration of stability and the conditions that affect the level at which performance stabilizes. Answers to these questions are certainly not available from the experimental literature of the past 100 years, and speculative generalizations are of little value until they are confirmed by definitive observations.

Longevity of Content as a Criterion of the Effectiveness of Education

Much of the content of primary school instruction (e.g., reading, writing, arithmetic) is rehearsed so frequently during the life of the individual that maintenance of content in the absence of rehearsal is not an important practical concern. However, this is not the case with regard to the content of high school and college courses.

The content of many high school and college courses may go unrehearsed, and the life span of unrehearsed content is not only of theoretical interest but is also an important criterion for evaluating educational programs. It would be difficult to argue for the value of a course if the knowledge acquired in that course is forgotten shortly after taking the final course examination. To be sure, teachers often claim that their courses are valuable because they teach their students how to think or because they promote the development of other mental faculties. Attempts to verify such claims have not been successful, however, and there is little evidence to support them.

To improve the effectiveness of education, we must evaluate the outcome of education, and we must base the evaluations on valid and reliable evidence of performance attributable to the program of instruction. The longevity of knowledge acquired in the classroom ought to be a key concern of educators and an important part of evaluating education. This is true even for content that is rarely used. Thus, training in cardiopulmonary resuscitation can be extremely valuable, but only if the knowledge is retained over long periods during which it may not be used.

Neither memory scholars nor educators have researched longevity of knowledge systematically. Although some investigators have measured retention of academic content during the past half century (e.g., Blizard, Carmody, & Holland, 1975; Blunt & Blizard, 1975; Cohen, 1976; Glasnapp, Poggio, & Ory, 1978; Sinclair, 1965; Spitzer, 1939; Smythe, Jutras, Bramwell, & Gardner, 1973; Wert, 1937), the retention intervals used were short and the research has not been programmatic, so that no large, relevant database is available.

It is known from laboratory findings that total practice and the spacing of practice affect the life span of episodic memory content, but the magnitude of these effects on the acquisition and retention of knowledge is not well established. Therefore, researchers can offer no guidelines to educators that would permit them to make informed decisions on the basis of expected trade-offs among the instructional costs of various schedules of acquisition and associated changes in the life span of knowledge. A relevant database and generalizations to guide curricular decisions constitute major potential contributions to education that increase the longevity of knowledge at an acceptable cost. The following discussion outlines and illustrates procedures for obtaining a relevant database.

Retention of Knowledge Acquired in a Statistics Course

To obtain information about the life span of the memory content taught in a basic statistics course, I obtained retention data for that course at three institutions: Ohio Wesleyan University, the Ohio State University, and Ben Gurion University, Israel. In each institution, undergraduates who had completed the basic statistics course during the previous academic year (and who had not used that content in other courses during the retention interval) were readministered the final examination. Table 1 shows the mean percentage of decline in scores from the first to the second administration for a retention interval of 9–12 months. Scores are reported separately for questions in recall and in recognition (four-alternative, multiple-choice) format. Recall scores declined by an average of 58% and recognition scores by 34%.

The decline of performance observed for the first year of retention exceeded substantially the decline obtained during the first year of retention for high school algebra content (33% to 65% vs. less than 10%). Algebra content may decline more slowly because prealgebra instruction in earlier grades is typically followed by a full academic year of algebra instruction in high school. This schedule provides for extended practice

TABLE 1

Mean Percentage of Decline of Performance on a Final Course Examination in Statistics, Readministered Within One Year

University	n	Recall score	Recognition score
Ohio Wesleyan	17	49	33
Ohio State	10		37
Ben Gurion	22	65	

and relearning of portions of the algebra content. No comparably extended acquisition period is common for the content of the basic statistics course. This content is usually taught during a single quarter or semester, and the content is minimally redundant with content covered in prior mathematics courses.

Retention Data Needed to Evaluate a Program of Instruction

The statistics retention data are based on a small sample size, but they illustrate the type of information needed to assess long-term retention of various content domains covered in a program of instruction. A standardized examination or the final course examination administered at the end of the course and readministered after intervals of 1–5 years are required. Some of the questions should be in recognition format, others in recall format, so that separate functions based on the two types of indicants can be established. If the retention data are longitudinal (i.e., the same individuals are tested more than once), data collection takes a long time, and the effects of repeated testing should be controlled by testing random portions of the content on successive administrations. If the data are cross-sectional, a multiple regression program must be used to equate successive data points with regard to the most important predictor variables (e.g., grades and aptitude test scores of participants). The cross-sectional approach saves time, but this advantage must be weighed against the disadvantages of imperfect control of variables pertaining to the amount and type of original learning. The problems of cross-sectional comparisons are particularly acute in domains in which major changes have been made in the content taught over successive years by various teachers using various textbooks and other instructional tools. The study of 50-year retention of high school algebra and plane geometry content (Bahrick & Hall, 1991a) dealt with these problems by limiting the retention test to subject matter that had remained constant between 1940 and 1990, as indicated by the New York State Regents Examinations from that period.

Residual knowledge and relative losses can be compared for various content areas by obtaining separate retention functions for each content area. Expert judgments of the relative importance of each domain of content must be obtained, and the comparative retention data together with the judgments of relative importance can then provide the basis for revising curricular strategies. The strategies involve methods of manipulating total exposure and the distribution of exposure of each content domain, with the objectives of achieving longer overall life spans of knowledge and life spans for individual domains that reflect their relative importance.

Several strategies to increase total exposure and distribution of exposure to content areas are discussed. In most cases, specific, quasi-experimental follow-up research is needed to assess and confirm the expected increments in the life span of knowledge. Available data may provide the basis for estimating some of these effects. The functions shown in Figure 2, for example, show the effects on the long-term retention of high school algebra content attributable to follow-up courses that review and apply the original content.

Methods of Implementing Reviews of Content

Strategies for increasing total exposure or distribution of exposure include ways of building mandatory reviews of previously covered content into the curriculum of spreading a given amount of instruction over a longer period of time. The simplest and most common method is the cumulative final course examination. Another common technique is the review of content covered in a prerequisite course during the first few sessions of follow-up courses. To enhance the value of such reviews, instructors should combine them with a reexamination of the reviewed content. Experimental findings supporting the substantial benefits of review examinations on the life span of knowledge were reported almost 70 years ago (Jones, 1923), but this excellent research was not continued and did not lead to significant changes of educational practice. Jones (1923) reported that a brief examination at the end of a lecture reduces retention losses on a delayed retest. The advantage attributable to the immediate test increases during the retention interval, and, after 2 months, the mean retention score of the previously tested group is twice as high as the mean score of the untested group.

Other, more elaborate ways of building reviews into a curriculum include cumulative, capstone review courses that may be combined with cumulative review examinations at the end of a program or at certain intermediate stages of a program as prerequisites to admission into the next stage of the program. An example of cumulative examinations that appear to have a significant effect on the life span of knowledge are the examinations traditionally required in European countries for obtaining a college-preparatory, secondary-school diploma. The examinations required for the "Abitur" in Germany cover several major topics, usually including languages and mathematics, and the outcome significantly affects admission decisions by German universities as well as other career opportunities.

The Distribution of Practice

The distribution of practice, independent of the total amount of practice, was first identified by Ebbinghaus (1885/1964) as an important variable affecting memory. This research issue has continued to generate research for more than a century (Bruce & Bahrick, 1992) and has generated important, clear-cut findings. Nevertheless, there is virtually no educational implementation of the laboratory findings on this issue, and this neglect was discussed in an important review by Dempster (1988). Although most of the laboratory findings are based on retention over short periods, Bahrick and Phelps (1987) showed that the benefits of distributed practice are evident after an interval of 8 years and appear to be enhanced, rather than diminished, during this long retention interval. In this study, college students learned and relearned 50 English–Spanish word pairs seven times to the same criterion. They were tested for recall and recognition 8 years later. The original relearning sessions were spaced either at 30-day intervals, at 1-day intervals, or all on the same day. Eight years later, participants who were trained at 30-day intervals recalled about twice as many words as those trained at 1-day intervals, and both of these groups retained more than the subjects who were trained and retrained on the same day.

All of the previously discussed strategies for reviewing and retesting educational content involve increments in the total amount of practice as well as the distribution of practice. Additional resources (teaching time and study time) must be committed to conduct reviews, prepare for cumulative examinations, and so forth. Thus, the benefits of increasing the life span of knowledge with these methods must be weighed against the cost of additional teaching and study time. Programmatic investigations performed in naturalistic settings can sort out the component contributions to increasing the longevity of knowledge by added practice, by distribution of practice, and by testing per se (Jones, 1923). Such research will yield cost–benefit data pertinent to various strategies.

However, a number of calendar and curricular changes may be made involving increments in the distribution of practice without requiring significant additional investments of teaching or study time. These changes maintain the total number of teaching hours devoted to a given content area, but distribute the hours over a longer period. On the simplest level, the 40 to 50 class hours typically devoted to a single college course can be distributed over 10 weeks, with 4 to 5 weekly meetings (typical for a quarter plan), or over 15 weeks, with 3 weekly meetings (typical for a semester plan). No published data are available to document that such a change yields benefits to long-term retention, but the findings of Bahrick and Phelps (1987) suggest that benefits are likely.

Much larger benefits can be predicted by extending the distribution of practice over a longer time period. If 40 to 50 hours of instruction are distributed over 2 years, by giving 1 hour of instruction every 2 weeks, there is good reason to believe that long-term retention would increase substantially. Evidence from Bahrick and Phelps (1987) suggests that slightly more study time may be needed to cover the same content under the extended schedule because forgetting between successive learning sessions requires somewhat more relearning when sessions are spaced far apart. However, even with an intersession interval of 30 days, the increments of study time are very small in relation to the extended life span attributable to the spacing effect.

Many programs of instruction are inherently sequential, and it is difficult to extend the time of acquisition of any portion of such content without exceeding the total instructional time available for the program. Thus, a basic statistics course is prerequisite to other methodology courses in most psychology programs, and extending instruction of basic statistics from 15 weeks to 1 or 2 years would make it difficult or impossible to complete the major in 2–3 years. Grading and evaluation practices are typically tied to 10- to 15-week intervals, and adjustments of these practices would also be necessary. In practice, increasing the life span of essential content may often require a combined strategy of relatively modest increases in the temporal distribution of original acquisition, together with required relearning of critical content, at intervals of 1 or 2 years after completing original acquisition. Such a combined strategy would not alter the total time needed to complete a program of instruction, and the benefits to long-term retention may be expected to outweigh the costs associated with relearning.

Maintenance of Access to Marginal Semantic Memory Content

Our findings (Bahrick & Hall, 1991b; Bahrick & Phelps, 1988), as well as those of others (Herrmann, Buschke, & Gall, 1987), indicate that large amounts of knowledge can be demonstrated reliably on tests of recognition, but are retrieved unreliably on tests of free or cued recall. Access to such targets on recall tests is unstable and fluctuates, depending on momentary context conditions. We refer to such knowledge as marginal. To illustrate, Herrmann et al. (1987) showed that subjects on a free-recall test list only approximately one third of the exemplars of semantic categories that they eventually list on subsequent tests and that they are able to identify on a recognition test. Our own findings regarding the retention of the Spanish language and of high school mathematics (Bahrick, 1984; Bahrick & Hall, 1991a) indicate that large amounts of knowledge that college students acquired in high school are marginal.

Instability of access impairs the usefulness of knowledge, for example, for professionals who may require prompt and dependable access to knowledge to carry out their responsibilities. Periodic interventions are necessary to assure maintenance of access to marginal knowledge, and it is of theoretical and practical importance to establish the effectiveness of various types and schedules of interventions designed to stabilize access to marginal knowledge.

The Method of Asymmetry of Fluctuations

Bahrick and Hall (1991b) have developed a method suitable for assessing the half-life of interventions designed to secure access to marginal semantic memory content. With this method, it is possible to determine the effectiveness of various types and schedules of interventions for various types of content and of subject populations.

Two apparently contradictory findings gave rise to our work on maintenance of access to marginal semantic memory content. The first finding is one that I have discussed previously. It showed that retention of knowledge systems not in active use may exhibit long periods of stability (i.e., no significant change in the overall retention level). The apparently contradictory finding is that these stable systems of knowledge include large amounts of information that is unstable on recall tests: Recall of individual target items fluctuates on successive tests depending on momentary context conditions. The apparent contradiction between the overall stability of recall performance and the large number of recall fluctuations for individual target items is resolved if it can be assumed that overall, the incidence of two kinds of fluctuations is equal. If, over any period of time, the number of targets changing from a nonaccessible to an accessible state equals the number of marginal targets changing in the opposite direction, overall recall level for that content would remain stable. It is a well-established fact, however (Payne, 1987), that repeated recall tests of the same target items yield asymmetrical, not symmetrical, fluctuations, with upward fluctuations more numerous than downward fluctuations. This asymmetry of fluctuation leads to an overall improvement of recall performance, a phenomenon described long ago by Ballard (1913) and more recently reviewed by Payne (1987). Hypermnesia has been observed for episodic as well as semantic content and has been explained as a testing effect, that is, as the cumulative result of repeatedly testing the same items (Roediger & Payne, 1982; Roediger, Payne, Gillespie, & Lean, 1982). This is the explanation of asymmetry that we adopt, and we use the degree of asymmetry of fluctuation as an indicant of the magnitude and the duration of testing or intervention effects.

With the method of asymmetry of fluctuations, we distinguish between two types of maintenance interventions: (a) Preventive maintenance interventions diminish the likelihood of losing access to currently accessible targets. (b) Corrective maintenance interventions reestablish access to currently inaccessible targets. In other words, preventive interventions prevent downward fluctuations of targets, and corrective interventions bring about upward fluctuations of targets.

The method requires repeated recall tests of the same targets belonging to a domain of knowledge that exhibits overall stability of recall level during the time of testing (the stability is verified by selecting two random samples of targets from the knowledge domain and by testing one sample at the beginning and the other at the end of the retention interval). If randomly selected target items that are not retested exhibit stability of overall recall, then asymmetry of fluctuation for retested items can be attributed to the testing or intervention effect. Asymmetry is complete immediately after testing (there are no downward fluctuations for items tested twice in immediate succession). The point in time at which asymmetry is reduced to 50% (half as many downward as upward fluctuations) designates the half-life of the testing or intervention effect. By using this method, we showed (Bahrick & Hall, 1991b) that the intervention associated with a single successful recall trial of marginal targets has a half-life in excess of one month.

Most or all of this effect appears to be a preventive maintenance effect, that is, the prevention of downward fluctuations of targets. If feedback or knowledge of results are given for some of the targets failed in the first recall test, asymmetry effects are augmented for these targets, and the augmentation serves as an estimate of the corrective maintenance effect.

The method of asymmetry of fluctuation permits a comparison of the half-life of intervention effects associated with a variety of preventive and corrective maintenance interventions. Thus, preventive maintenance effects can be compared for one, two, or more tests separated by various time intervals; and corrective maintenance effects can be determined for various types and schedules of feedback.

Programmatic research using the method of asymmetry of fluctuation will yield generalizations regarding the maintenance of access to marginal semantic memory content. Based on such generalizations, efficient schedules of testing and rehearsal can be developed. These will ensure continuing access to marginal content on the basis of relatively low investments of testing and rehearsal time.

It is also likely that continuing maintenance interventions will gradually change marginal content into permastore content. This process occurs if repeated interventions of the same type yield progressively longer lasting maintenance effects. Eventually, downward fluctuations no longer occur, and there is no need for further maintenance interventions. The method of asymmetry of fluctuation will make it possible to investigate this important transition.

Programmatic investigations of the effects of distribution of practice, of total practice, of forced review, and of periodic maintenance interventions are a necessary basis for cost-effective extensions in the life span of knowledge and in the accessibility of marginal knowledge. Cost-effective extensions in the life span of knowledge constitute a major potential contribution of memory scholarship to education.

References

Bahrick, H. P. (1984). Semantic memory content in permastore: Fifty years of memory for Spanish learned in school. *Journal of Experimental Psychology: General, 113*, 1–29.

Bahrick, H. P. (1989). The laboratory and the ecology: Supplementary sources of data for memory research. In L. Poon, D. Rubin, & B. Wilson (Eds.), *Everyday cognition in adulthood and late life* (pp. 73–83). Hillsdale, NJ: Erlbaum.

Bahrick, H. P. (1992). Stabilized memory of unrehearsed knowledge. *Journal of Experimental Psychology: General, 121*, 112–113.

Bahrick, H. P., Bahrick, P. O., & Wittlinger, R. P. (1975). Fifty years of memory for names and faces: A cross-sectional approach. *Journal of Experimental Psychology: General, 104*, 54–75.

Bahrick, H. P., & Hall, L. K. (1991a). Lifetime maintenance of high school mathematics content. *Journal of Experimental Psychology: General, 120*, 20–33.

Bahrick, H. P., & Hall, L. K. (1991b). Preventive and corrective maintenance of access to knowledge. *Applied Cognitive Psychology, 5*, 1–18.

Bahrick, H. P., & Phelps, E. (1987). Retention of Spanish vocabulary over eight years. *Journal of Experimental Psychology: Learning, Memory, and Cognition, 13*, 344–349.

Bahrick, H. P., & Phelps, E. (1988). The maintenance of marginal knowledge. In U. Neisser & E. Winograd (Eds.), *Remembering reconsidered: Ecological and traditional approaches to the study of memory* (pp. 178–192). Cambridge, England: Cambridge University Press.

Ballard, P. B. (1913). Oblivescence and reminiscence. *British Journal of Psychology: Monograph Supplement, 1*, 1–82.

Banaji, M. R., & Crowder, R. G. (1989). The bankruptcy of everyday memory. *American Psychologist, 44*, 1185–1193.

Banaji, M. R., & Crowder, R. G. (1991). Some everyday thoughts on ecologically valid methods. *American Psychologist, 46*, 78–79.

Blizard, P. J., Carmody, J. J., & Holland, R. A. (1975). Medical students' retention of knowledge of physics and chemistry on entry to a course in physiology. *Journal of Medical Education, 9*, 249–254.

Blunt, M. J., & Blizard, P. J. (1975). Recall and retrieval of anatomical knowledge. *British Journal of Medical Education, 9*, 255–263.

Bruce, D. (1991). Mechanistic and functional explanations of memory. *American Psychologist, 46*, 46–48.

Bruce, D., & Bahrick, H. P. (1992). Perceptions of past research. *American Psychologist, 47*, 319–328.

Ceci, S. J., & Bronfenbrenner, U. (1991). On the demise of everyday memory: "The rumors of my death are much exaggerated" (Mark Twain). *American Psychologist, 46*, 27–31.

Cohen, A. D. (1976). The Culver City Spanish immersion program: How does summer recess affect Spanish speaking ability. *Language Learning, 24*, 55–68.

Conway, M. A. (1991). In defense of everyday memory. *American Psychologist, 46*, 19–26.

Conway, M. A., Cohen, G., & Stanhope, N. (1991). On the very long-term retention of knowledge acquired through formal education: Twelve years of cognitive psychology. *Journal of Experimental Psychology: General, 120*, 358–372.

Dempster, F. N. (1988). The spacing effect: A case study in the failure to apply the results of psychological research. *American Psychologist, 43*, 627–634.

Ebbinghaus, H. E. (1964). *Memory: A contribution to experimental psychology.* New York: Dover. (Original work published 1885)

Glasnapp, D. R., Poggio, J. P., & Ory, J. C. (1978). End-of-course and long-term retention outcomes for mastery and nonmastery learning paradigms. *Psychology in the Schools, 15*, 595–603.

Herbart, J. F. (1982). Psychologie als Wissenschaft neu gegrundet auf Erfahrung, Metaphysik und Mathematic [Psychology as science newly founded in experience, metaphysics and mathematics]. In K. Kehrbach (Ed.), *J. F. Herbart's samtliche Werke* (Part 1 is in Vol. 5, pp. 177–434; Part 2 is in Vol. 6, pp. 1–340). Langensalze: H. Beyer & Sohne. (Original work published 1824)

Herrmann, D. J., Buschke, H., & Gall, M. B. (1987). Improving retrieval. *Applied Cognitive Psychology, 1*, 27–33.

Hintzman, D. L. (1993). 25 years of learning and memory: Was the cognitive revolution a mistake? In D. E. Meyer & S. Kornblum (Eds.), *Attention and performance* (Vol. 14, pp. 359–391). Cambridge, MA: MIT Press.

Jones, H. E. (1923). Experimental studies of college teaching: The effect of examination on permanence of learning. *Archives of Psychology, 68*, 1–70.

Kintsch, W. (1974). *The representation of meaning in memory.* New York: Wiley.

Klatzky, R. L. (1991). Let's be friends. *American Psychologist, 46*, 43–45.

Loftus, E. F. (1991). The glitter of everyday memory . . . and the gold. *American Psychologist, 46*, 16–18.

Neisser, U. (1978). Memory: What are the important questions? In M. M. Gruneberg, P. E. Morris, & R. N. Sykes (Eds.), *Practical aspects of memory* (pp. 3–20). San Diego, CA: Academic Press.

Neisser, U. (1991). A case of misplaced nostalgia. *American Psychologist, 46*, 34–36.

Payne, D. G. (1987). Hypermnesia and reminiscence in recall: An historical and empirical review. *Psychological Bulletin, 101*, 5–27.

Roediger, H. L., III, & Payne, D. G. (1982). Hypermnesia: The role of repeated testing. *Journal of Experimental Psychology: Learning, Memory, and Cognition, 8*, 66–72.

Roediger, H. L., III, Payne, D. G., Gillespie, G. L., & Lean, D. S. (1982). Hypermnesia as determined by level of recall. *Journal of Verbal Learning and Verbal Behavior, 21*, 635–655.

Sinclair, D. (1965). An experiment in the teaching of anatomy. *British Journal of Medical Education, 40*, 401–413.

Smythe, P. C., Jutras, G. C., Bramwell, J. R., & Gardner, R. C. (1973). Second language retention over varying intervals. *Modern Language Journal, 57*, 400–405.

Spitzer, H. (1939). Studies in retention. *Journal of Educational Psychology, 30*, 641–656.

Wert, J. E. (1937). Twin examination assumptions. *Journal of Higher Education, 8*, 136–140.

CHAPTER 3

Cognition and Emotion: Theories, Implications, and Educational Applications

Henry C. Ellis, Larry J. Varner, and Andrew S. Becker

T his chapter describes some of the principal theories of cognition and emotion, discusses their implications and applications, and outlines some of the major coping strategies that have been proposed to deal with emotional stress in the context of classroom learning. There has been considerable growth in research on cognition–emotion relationships in recent years, and this research has been summarized in several sources (e.g., Ellis & Ashbrook, 1988, 1989; Ellis & Hertel, 1993; Fiedler & Forgas, 1988; Kuiken, 1989, 1991; Williams, Watts, MacLeod, & Mathews, 1988). Since about 1975, research in cognition and emotion has accelerated at a rapid rate. In addition, a new journal titled *Cognition and Emotion*, devoted to the study of relations among emotional states and to the full range of cognitive processes studied by psychologists, appeared in 1987. Ellis and Ashbrook (1989) noted that, although this research area had a much earlier history of activity, it lay relatively dormant until the 1970s

We are grateful for the careful and thoughtful reviews of an earlier draft of this chapter by Douglas Nelson and Charles Spielberger.

for several reasons (Ellis & Hunt, 1993). One was that few cognitive psychologists evidenced much interest in affect until the mid-1970s. Another factor was the preoccupation of cognitive psychologists with understanding cognitive processes per se, that is, cognitive processes in the absence of affect. Perhaps another factor was the absence of extensive theory and the lack of adequate methods for studying cognition and emotion. Nevertheless, the scene has changed dramatically since the 1970s, and we describe some of these developments.

The objectives of this chapter are threefold. We first describe some of the principal theories of cognition and emotion. These theories have generally developed within cognitive psychology, personality theory, and clinical psychology, and we have selected theories that have implications for and applications to the learning of mathematics and science topics. Then, we review selectively some of the classroom learning literature with findings relevant to these theories. Finally, we examine some of the coping strategies proposed to handle stress, particularly in educational settings, and note the implications that several theories have for the strategies designed to deal with stress and emotional states. More generally, our overall approach is to provide a framework for considering the role of affect on cognitive processes and to consider the implications of this framework for educational performance.

Theories of Cognition and Emotion

A variety of theories exist concerning the effects of emotional states on cognitive processes, varying from restricted theories dealing with specific phenomena such as processing initiative (Hertel & Hardin, 1990) to more general theories dealing with anxiety and performance (e.g., Eysenck, 1982), resource allocation and cognitive interference (e.g., Ellis & Ashbrook, 1988), and frameworks such as schema theory (e.g., Beck, 1967). The majority of these theories have focused either on the role of anxiety as it influences cognitive processes, or on depression and general mood states as they affect cognitive processes and performance. In both cases, that is, in theories of anxiety and of depression, there have been extensive attempts not only to describe their effects on performance measures, but also to understand the range of cognitive processes influenced by these emotional states. Broad theories of behavior that relate motivational–emotional states to performance (e.g., Easterbrook, 1959) have also had a long history in psychology. We do not, however, consider these more overarching theories, despite their historical interest, principally because they have relatively little influence either in current theoretical work or in applications to educational issues.

Theories of anxiety and performance have been developed extensively; have wide relevance to issues in clinical, personality, social, and cognitive psychology; and have been used to account for findings in classroom studies, particularly with regard to test anxiety (e.g., Carver, Scheier, & Klahr, 1987; Geen, 1987; Ramirez & Dockweiler, 1985; Schwarzer, van der Ploeg, & Spielberger, 1987). Theories of depression and related emotional states have also been developed extensively and also have wide relevance to clinical, personality, social, and cognitive psychology. However, it would appear that such theories, although applicable to educational settings, have had less frequent application in efforts to understand classroom performance. We contend, however, that depression and related emotional states are equally relevant in understanding performance in classroom settings because of the generally recognized importance of emotional mood states in a variety of settings (e.g., Ellis & Ashbrook, 1988; Fiedler & Forgas, 1988; Kuiken, 1991; Williams et al., 1988). Therefore, we review selected theories that have either direct or potential application to the understanding of performance in the classroom. It is not our intent to be encyclopedic; instead, we examine six theories concerning either anxiety or depression and related mood states, theories that have current or potential usefulness in understanding classroom learning. We examine the following theories: (a) anxiety and attentional interference, (b) arousal and motivational theories of anxiety, (c) processing efficiency theory and anxiety, (d) resource allocation theory and cognitive interference, (e) network theory, and (f) schema theory. In addition, we consider briefly resistance to distraction, processing initiative, and inhibitory processes as three potentially useful ideas.

Anxiety and Attentional Interference

Theories of anxiety and performance frequently distinguish between worry and emotionality as factors that can impair cognitive activities.

Worry refers simply to the intruding thoughts that reflect self-concern, doubt, or other negative events. *Emotionality*, in turn, refers to the heightened arousal state of the individual. Theories of anxiety have focused on the aspects of worry that involve self-preoccupation and thoughts about evaluation and feelings of inadequacy. A major theory regarding the effects of anxiety on performance is that of Sarason (1984, 1988), who explained the effects of anxiety on performance by way of its effects on attention given to the task at hand. Sarason proposed that the production of worrisome thoughts interferes with or even precludes, in extreme cases, the allocation of attention to task-relevant information, thus reducing essential cognitive resources that would otherwise be available for processing the criterion task. The reduction of attentional resources thus leads to impaired performance. This assumption of attentional interference, that is, the reduced allocation of attentional resources, is a major assumption of Ellis's resource allocation model of depression (Ellis, 1984, 1985, 1990; Ellis & Ashbrook, 1988; Ellis, Thomas, & Rodriguez, 1984), and it attests to one important similarity between the theories of anxiety and depression.

Sarason's (1984) theoretical position leads to two predictions. One is that high-anxious individuals will generally show poorer performance than low-anxious individuals. The other is that anxiety will produce greater deficits with more difficult, demanding tasks. Again, parallel predictions are made for depression and, more generally, for all disruptive mood states in Ellis's resource allocation model. A number of studies support Sarason's first prediction (e.g., Deffenbacher, 1978); however, not all studies show cognitive impairment due to heightened anxiety. Similarly, a number of studies support the anxiety and task difficulty interaction; however, there are some problems in the conceptualization of task difficulty that make the interpretation of this interaction complicated (cf. Ellis et al., 1984). Although attentional interference is seen as an important mechanism in understanding performance deficits associated with anxiety, Eysenck and Calvo (1992) pointed to the importance of additional mechanisms, which are described in the next two theories.

Arousal and Motivational Theories of Anxiety

A variety of theories of cognition and emotion have implicated an arousal component. For example, Clark, Milberg, and Erber (1988) proposed that arousal can cue information stored in memory with a similar level of arousal. This general idea has important implications for education in that arousal can reinstate previous emotional states present during the learning of anxiety-evoking information. Similarly, Humphreys and Revelle (1984) and Revelle and Loftus (1990) proposed a general theory of anxiety in which arousal is a key component. According to their theory, state anxiety has a cognitive component (worry) that, like Sarason's (1984) theory, produces a reduction in the allocation of resources to some criterion task. In addition, state anxiety has an arousal component that leads to heightened alertness, vigor, and activation. Characteristics of the task are also important, particularly if the task makes heavy demands on short-term memory (STM). Performance on such tasks is assumed to be impaired by anxiety. The worry component is thought to interfere with both STM tasks and those involving more sustained processing, whereas the arousal component is conceived as playing a more complex role. Arousal is thought to interfere with STM tasks but, in contrast, is assumed to facilitate tasks involving more sustained performance. The important feature of this theory is that performance deficits produced by worry can be counteracted by high arousal, thus calling attention to the potential role of compensatory effort in preventing performance deficits in anxious subjects. A recent study by Hertel and Rude (1991) pointed out the role of focused attention as a process that can overcome performance decrements caused by depression, a finding having clear therapeutic as well as educational implications (Ellis, 1991). Humphreys and Revelle (1984) assumed that similar compensatory activities can occur in anxious individuals.

Eysenck (1982) also proposed an arousal theory designed to account for the effects of anxiety on performance. He argued that individuals high in trait anxiety and in state anxiety have less attentional capacity available in working memory for adequate performance of some criterion task than those low in trait anxiety. This occurs because worry, self-concern, or other task-irrelevant cognitive activities tie up some portion of the available limited capacity. The principal prediction from this theory is that detrimental effects caused by anxiety should be greater with tasks that make heavy demands on attentional capacity.

Processing Efficiency and Anxiety

Most recently, Eysenck and Calvo (1992) expanded this theory into a version that they called *processing efficiency theory*. Processing efficiency theory makes two main predictions regarding the effects of worry and anxiety. One is that worry brings about a reduction in the storage and processing capacity of the working memory system, an assumption of his earlier theory. The second is that worry brings about an increment in on-task effort and activities designed to improve performance and to avoid aversive consequences. This second prediction also calls attention to the motivational aspects of worry and, in this regard, is similar to Sarason's (1984) position. To escape from a state of apprehension associated with worry and to avoid the aversive consequences of failure, anxious individuals tend to allocate greater effort (attentional resources) and to initiate more strategies likely to produce acceptable performance.

Eysenck and Calvo (1992) made an important distinction between performance effectiveness and processing efficiency. *Performance effectiveness* refers to the quality of task performance, whereas *processing efficiency* refers to the relationship between the effectiveness of performance and the effort or resources deployed in performing the criterion task. Theoretically, Eysenck and Calvo defined processing efficiency as performance effectiveness divided by effort. The importance of this distinction, as seen by Eysenck and Calvo, is that anxiety may have different effects on performance quality and on processing efficiency. For example, the detrimental effects of anxiety on performance effectiveness may be less than on processing efficiency. If a person allocates greater resources to a task when under high anxiety, performance may fail to suffer or show only small impairment, whereas the processing efficiency of the person may show considerable attenuation because of the greater effort allocated to the task.

Resource Allocation Theory and Cognitive Interference

The next type of theory stems from capacity models of attention and proposes that performance is constrained by the amount of capacity available to handle the demands of some current cognitive task (e.g.,

Kahneman, 1973). Adopting the concept of capacity allocation, Ellis (1984, 1985; Ellis & Ashbrook, 1988; Ellis, Thomas, McFarland, & Lane, 1985; Ellis et al., 1984) developed a resource allocation model in which many of the findings regarding the effects of depressed mood states on cognitive processes might be explained. The most complete description is seen in Ellis and Ashbrook (1988), and the model makes five major assumptions: (a) Emotional mood states produce their effects on cognitive activities by regulating the amount of capacity available that can be allocated to a given criterion task. Specifically, a depressed mood state reduces the amount of capacity available that can be allocated to a given criterion task because the depressed mood state preempts or ties up some of the capacity that would otherwise be allocated to the criterion task. (b) The encoding of information usually requires some cognitive effort or capacity to be allocated to the criterion task. Although the idea of automatic processing is recognized, it is contended that the majority of everyday tasks that we deem important require at least minimal allocation of cognitive resources. (c) Retention performance may be positively correlated with the amount of cognitive effort allocated to the criterion task. This assumption has received support in several studies (e.g., Ellis et al., 1984; Jacoby, 1978; McFarland, Frey, & Rhodes, 1980; O'Brien & Myers, 1985; Swanson, 1984; Tyler, Hertel, McCallum, & Ellis, 1979). At present, this assumption implies only a correlation between effort and memory and does not assume a causal relationship; moreover, the mechanisms underlying this relationship are yet to be understood. This model makes several predictions, one of which contends that depressed mood subjects will perform more poorly on criterion tasks as encoding effort demands increase because fewer resources are available. There is empirical support for this prediction from both experimental and clinical studies (Ellis & Ashbrook, 1988).

The last two assumptions of the model deal with the role of irrelevant thoughts and their interference with cognitive processes. It is proposed that (d) the production of irrelevant thoughts increases under emotional duress and (e) these irrelevant, distracting thoughts, in turn, interfere with the ability to encode, organize, and retrieve information, thus leading to poorer recall. These assumptions lead to the prediction that emotional states produce their effects on memory and other cognitive processes not by way of emotion per se, but by way of distracting, competing, irrelevant thoughts that interfere with successful performance of cognitive tasks. From the perspective of classical learning theory, emotion can impair cognitive processes because of the disruptive competing-response thought activities that occur.

Prevalent in several theoretical accounts of the effects of mood states on memory is the idea that mood states produce a prevailing pattern of thoughts that, in turn, influence performance in a variety of cognitive tasks (e.g., Beck, Rush, Shaw, & Emery, 1979; Ellis & Ashbrook, 1988; Ingram, 1984). An explicit assumption in these theories has been that depressed mood state may have its effect on memory and cognition by way of negative or unfavorable self-thoughts that interfere with performance on some cognitive task. As a first step in understanding the possible relationship between mood states and their resultant thoughts, Ellis, Seibert, and Herbert (1990) examined the effects of the induction of a depressed mood on the production of unfavorable thoughts and found that subjects who were depressed produced far more negative, self-damaging reports than neutral controls, even though both groups (depressed and neutral) reported the same total number of thoughts. Thus, the effects were not the product of differential response bias in output. Second, they found that depressed mood subjects reported more unfavorable thoughts, even when given a memory task, and that they performed poorer at recall.

Continuing this line of research, Seibert and Ellis (1991) examined the relationship between irrelevant, distracting thoughts and memory. Subjects were asked either to produce all their thoughts (which were tape-recorded) in a mood–memory experiment, or were asked to list their thoughts at the end of a mood–memory experiment. Subjects were asked to rate their reported thoughts on a 5-point scale as being relevant or irrelevant to the criterion memory task, a procedure that allowed an examination of the relationship between the production of irrelevant thoughts and recall of the criterion memory task. Using either concurrent verbalization or thought listing, the results were strikingly clear. The production of irrelevant thoughts, as judged by subjects, was strongly negatively correlated with recall ($-.72$ and $-.67$, respectively). Although these findings are correlations, they strongly support the view that mood effects on memory and cognition are the results of thoughts, that is, of cognitive activation that intervenes between the mood state and performance on the criterion task. This view makes the important argument that it is the *cognitive consequence* of emotional mood states that has an impact on memory, as distinct from the emotional or affective state itself. More generally, these experiments show that it is possible to examine thoughts as such directly and to relate them to other cognitive processes.

Network Theory

Another view of the role of emotion in memory and cognitive processes is that of network theory. Bower's (1981) semantic network theory assumes that emotional states are represented as nodes in semantic memory and that emotions such as joy, depression, or fear are represented by a specific node or unit including related aspects of each emotion. In addition, each emotion node is linked with propositions that describe events from a person's life during which that emotion was aroused. Emotion nodes can be activated by many events and are subject to the spread of activation as are other memory representations. The network theory has been used to interpret several mood effects, especially mood-congruent effects and state-dependent effects (Blaney, 1986). Mood-congruent effects are seen as the result of activation of emotion nodes by some emotional event that is consistent with a person's network of emotional memories. State-dependent events were not found to be robust by Bower and Mayer (1989); however, Eich (1989) reported a number of studies with positive results.

Network theory may ultimately have some potential for application in understanding the role of affect in classroom learning. However, there appears to have been little attempt to exploit this conceptual approach in this setting. Nevertheless, we note it because of its possible potential for understanding events like mathematics anxiety in which mathematics events (e.g., equations) may serve to reinstate negative emotional states associated with earlier aversive experiences in learning mathematics.

Schema Theory

Schema theory bears some relationship to network theory and can also account for some of the effects of various emotional states on cognitive processes. With regard to emotional states, schema theory contends that a person's prevailing mood state functions as a conceptual structure (schema) for processing and organizing incoming information as well as for guiding the retrieval process (e.g., Beck, 1967; Kuiper, MacDonald, & Derry, 1983). With regard to depression, the basic assumption is that a depressed person has a predisposition toward negative events that are organized into a depressed schema. The mood-as-schema conceptual approach is neither inconsistent with a resource allocation approach nor does it necessarily predict different performance; it simply stems from a different conceptual base (Ellis & Hunt, 1993). Several empirical findings are consistent with this attempt to account for cognition–emotion relationships, the most salient of which are mood-congruent effects in memory. In general, these studies have reported that subjects in a specific mood (e.g., depression) are more likely to recall depressed mood content from word lists or stories than are happy subjects and vice versa. This issue was studied extensively by Teasdale and Fogarty (1979) and by Teasdale and Russell (1983). With regard to emotional states, schema theory has been used most extensively as a framework in cognitive theories of depression and in psychotherapy. The leader in this development is Beck (1967), who argued for the importance of understanding the cognitive basis of depression.

Schema theory is certainly applicable to the understanding of classroom performance, if mood states of persons are viewed as part of a prevailing schema that individual learners bring to the classroom. A student with a prevailing negative schema about learning, or with a prevailing schema that expects failure or other aversive outcomes, may certainly be expected to show poorer performance. With such students, the teacher has a major challenge in trying to provide experiences that will modify and change the prevailing negative schema. This, of course, is no simple task and may require a number of positive learning experiences to overcome this negative schema.

Other Theoretical Ideas

Finally, we briefly focus on three additional processes that are important aspects of understanding the effects of emotional states on cognition. These processes concern the ease of distraction that individuals display in learning situations (Ellis & Franklin, 1983), processing initiative (Hertel & Hardin, 1990), and inhibitory mechanisms in learning and memory (Hasher & Zacks, 1988).

The importance of distraction in learning settings was shown by Ellis and Franklin (1983). Their study involved a comparison of externals and internals on the Rotter scale with regard to memory for categorized word lists. In the typical situation, externals remembered as well as internals in free recall. However, when distracting color stimuli were presented with the word list, externals were significantly poorer in list recall and were more distracted by the color information. This study suggests that predisposition to distraction may be an important feature in ordinary learning and is attested to by the importance of the concept of attention deficits in classroom learning. Moreover, the evidence suggests that heightened emotional states are more likely to predispose an individual to distraction.

A second concept, introduced by Hertel and Hardin (1990), proposed the interesting idea that depression may inhibit the *initiation* of cognitive activities. They assume that once cognitive processes are initiated, depression may have little or no effect on performance. The impairment is thus in the initiation of activity, not in sustained cognitive processes once activated.

The third concept that may have potential usefulness is that of inhibition in working memory. Neumann (1987) pointed out that the effective operation of working memory depends on inhibitory mechanisms that serve to limit what enters working memory to information that is pertinent to understanding a message. Hasher and Zacks (1988) adopted this idea in the understanding of how older adults comprehend information, which focuses on effective inhibitory control. As they noted, a breakdown in the efficiency of inhibitory mechanisms may have very important consequences. Indeed, such a breakdown may lead to cross talk among simultaneously active messages, preventing both a clear understanding of the message and the production of an adequate response. In extreme cases, this is reflected in such behaviors as attention deficit disorder. Given the potential importance of inhibitory mechanisms in working memory, this concept also deserves greater attention in the understanding of classroom performance. A summary of the principal concepts used in theories of cognition and emotion is given in Table 1.

Selected Review of Literature

Six theories identifying the effects of emotional states on cognitive processes were described in the previous section. Many of them have influenced the way in which studies on mathematics and science performance have evolved and the way in which findings have been interpreted. This section examines selected literature on the role of emotional states in

TABLE 1

Summary of Principal Concepts in Theories of Cognition and Emotion

Concepts	Authors
Attentional interference	Sarason (1984, 1988)
	Ellis and Ashbrook (1988)
Arousal	Clark, Milberg, and Erber (1988)
	Humphreys and Revelle (1984)
	Revelle and Loftus (1990)
	Eysenck (1982)
Processing efficiency	Eysenck and Calvo (1992)
Resource allocation	Ellis and Ashbrook (1988)
	Sarason (1984, 1988)
Irrelevant processing and	Ellis and Ashbrook (1988)
cognitive interference	Seibert and Ellis (1991)
	Sarason (1988)
	Ingram (1984)
	Ellis, Seibert, and Herbert (1990)
Emotion nodes in networks	Bower (1981)
Mood-as-schema	Beck (1967)
	Kuiper, MacDonald, and Derry (1983)
Resistance to distraction	Ellis and Franklin (1983)
Processing initiative	Hertel and Hardin (1990)
	Ellis (1990)
Inhibitory mechanisms	Hasher and Zacks (1988)

classroom performance in mathematics and science in terms of two common theoretical frameworks: attentional theory and arousal theory. Attentional theories, including Sarason's (1984, 1988) anxiety theory and Ellis's (1985; Ellis & Ashbrook, 1988; Ellis et al., 1985, 1984) resource allocation model, use the notion of a limited attentional capacity that can be disrupted by cognitions associated with emotions. Arousal theories, including Humphreys and Revelle's (1984) and Revelle and Loftus's (1990) theories, state that arousal associated with emotions plays a complex role, thus interfering with performance in STM tasks or facilitating performance in more sustained tasks. Although these two theoretical frameworks provide different perspectives for studying performance in mathematics and science, it will been seen that the research from these two areas converge on the common finding that classroom performance is affected by both worry and emotionality.

Research Using Attentional Theories

Most of the mathematics and science educational studies are based principally on attentional theories of emotion. These theories include Sarason's (1984, 1988) attention interference theory and Ellis's (1985; Ellis & Ashbrook, 1988; Ellis et al., 1984, 1985) resource allocation theory. Attentional theories are based on three general assumptions: (a) that humans possess a limited-capacity attentional system, (b) that cognitions associated with emotions disrupt the allocation of the attentional capacity and/or reduce the total amount of capacity that can be deployed to a cognitive task, and (c) that worry and other emotionally laden activities can increase the production of irrelevant thoughts that compete with the productive thoughts necessary for successful completion of the criterion task. Many studies using attention as their theoretical basis identify two components of emotions that can affect mathematics and science performance: worry and emotionality (Herman, 1990; Morris & Liebert, 1969, 1970; Sarason, 1984, 1988). *Worry* is defined as cognitive concern about one's performance, and *emotionality* as autonomic reactions that occur in response to the emotional stress of a situation. These studies have frequently assumed that it was interference from intrusive cognitions that decreased mathematics and science performance rather than the arousal produced by the testing situation. However, it will be seen that arousal also affects performance.

Liebert and Morris (1967) were among the first to propose the distinction between the worry and emotionality components of anxiety. Their idea was that worry varied intensely with expectancy and performance in test situations, whereas emotionality was unrelated to either expectancy or performance. In two articles, Morris and Liebert (1969, 1970) tested these hypotheses. They measured worry and emotionality by using a portion of Mandler and Sarason's (1952) Test Anxiety Questionnaire that referred to the subject's present feelings; in addition, they measured the subject's expected and actual scores on different classroom tests. They found, in agreement with their hypotheses, that worry correlated negatively with both the subject's expected and actual scores on all the tests; in contrast, emotionality was not related to either the subject's expected or actual scores. These findings indicate that the cognitive component of anxiety was interfering with attentional focus in classroom performance.

Several studies have tried to address the question of how cognitions that accompany worry affect performance on mathematics and science tests (Blankstein, Flett, Boase, & Toner, 1990; Herman, 1990; Hunsley, 1987; Kent & Jambunathan, 1989; Sarason, 1984). They have looked directly at how cognitions vary with performance in test situations by exploring the degree to which cognitions interfere with performance. For example, Sarason (1984) examined this by measuring worry (using Mandler & Sarason's [1952] Test Anxiety Questionnaire), cognitive interference (using Sarason & Stoops's [1978] Cognitive Interference Questionnaire), and test anxiety (using Sarason's [1978] Test Anxiety Scale). He found that worry and cognitive interference were positively correlated and that both were negatively correlated with performance. In addition, he found a positive correlation between test-irrelevant thinking and cognitive interference: The former also correlated negatively with performance. More generally, Sarason's findings support the idea that performance denigration is due to intrusive, irrelevant thoughts that occupy or direct attention away from the task at hand.

The issue of what types of thoughts that interfere with performance are present at test time was addressed by Blankenstein et al. (1990), who examined the content of the thoughts that subjects were experiencing during testing periods by way of a thought listing procedure. Subjects were asked to list their thoughts as they occurred during a test period and were instructed to be as spontaneous as possible. In addition, the subjects completed the Cognitive Interference Questionnaire (Sarason & Stoops, 1978), the Test Anxiety Scale (Sarason, 1978), and a physiological arousal scale designed by the authors. They found that test anxiety was positively correlated with negative self-referential thoughts listed. As with Sarason (1984), test anxiety was positively correlated with cognitive interference but not with physiological arousal. Interestingly, the level of test anxiety did not affect the total number of thoughts listed by the subjects; only the ratio of negative to positive self-referential thoughts was affected. This finding is in accordance with Ellis et al. (1990), who found that although depression did affect the quality of thoughts listed, with the level of depression positively correlated with the number of negative thoughts reported, the overall quantity of thoughts listed was unaffected. More generally, the Blankstein et al. (1990) study reveals that it is the kind of thoughts (i.e., quality) that a person has that interferes with cognitive processes and with the allocation of attention to the task.

Also of interest is that several of these studies have examined the effect that emotion has on a subject's expectancies about performance. Blankstein et al. (1990), Hunsley (1987), and Morris and Liebert (1970) have each looked at the effect that anxiety has on subjects' ratings of their upcoming performance on a variety of tests. All observed similar effects of anxiety on subjects' verbalized expectancies during actual classroom test performance. Worry correlated negatively with expectancies of performance on the upcoming tasks, whereas emotionality was unrelated to the expectancies. This finding is important because it is possible that the expectation of poor performance primes the subjects to experience the negative cognitions, which, in turn, affects performance on the subsequent test. Treating these initial negative expectancies may alleviate much of the anxiety experienced during the actual test period.

In summary, studies of the attentional aspect of emotions and their effect on mathematics and science performance provide three important findings. First, emotions about an upcoming test may be initiated well before the test, thus affecting subjects' expectancies about performance. Second, it is possible that persons' negative expectancies about their performance can persist during the test of classroom performance. Third, it is the quality of the cognitions, that is, the negative self-referential thoughts, during testing that interfere with performance.

Research Using Arousal Theories

The second area of mathematics and science educational research to be examined concerns those studies that concentrate on the arousal component of anxiety. Whereas the earlier literature examined the worry component of anxiety (i.e., cognitions interfering with performance), the arousal literature examines the emotionality component of anxiety (i.e., physiological arousal associated with performance). Emotionality in this body of literature is also related closely to state anxiety as defined by Spielberger, Gorsuch, and Lushene (1970). According to the previously discussed theories, state anxiety includes a cognitive component, but it is characterized by a feeling of apprehension and heightened autonomic arousal (Humphreys & Revelle, 1984; Revelle & Loftus, 1990; Sewell, Farley, & Sewell, 1983). Although state anxiety includes a cognitive component, it does not endure as long as worry; Morris and Liebert (1969, 1970) noted that emotionality or state anxiety rises sharply immediately before a test and decreases sharply immediately after a test. These studies examine how much state anxiety and/or emotionality affect performance in addition to trait anxiety and/or worry.

A study by Sewell et al. (1983) examined how state anxiety and trait anxiety, measured by the State-Trait Anxiety Inventory (Spielberger et al., 1970), and cognitive style, measured by the Group Embedded Figures Test (Witkin, Oltman, Raskin, & Karp, 1971), affect mathematics achievement on three tests during a semester. They found that only state anxiety was negatively correlated with mathematics performance. From this, they concluded that mathematics performance was directly related to state anxiety at the time of testing but was not directly affected by one's proneness to anxiety (i.e., trait anxiety).

Spielberger, Anton, and Bedell (1976) discussed the relationship between the cognitive and emotional components of test anxiety. They assumed that high levels of state anxiety, which correspond roughly to emotionality as defined by Liebert and Morris (1967), activate task-irrelevant worry responses and that these responses distract the individual from effective task performance. They also concluded that state anxiety may activate task-related error tendencies that, in turn, compete with correct responses. These assumptions are in contrast with Liebert and Morris (1967), who assumed that it is worry that affects performance on intellectual cognitive tasks and that worry about performance is what produces the state anxiety and increased arousal.

An important finding by Hodapp (1982) may provide a resolution to these contrasting approaches to studying test anxiety. He focused on how emotionality and worry contributed to achievement in testing situations. His initial finding, congruent with the attentional literature, was that worry negatively correlated with achievement, whereas emotionality was unrelated to achievement. He then studied how all the variables explored (worry, emotionality, state anxiety, and trait anxiety) were related to achievement through their predictive capacity by examining the covariance of the variables and by developing a path analytic model. He found that both emotionality and worry reliably predicted achievement and that trait anxiety influenced achievement by influencing both emotionality and worry. Hodapp (1982) concluded that although worry appears to be the major factor contributing to test anxiety, emotionality shows a relationship to achievement as well. He also concluded that the kind of task and degree of difficulty may influence the strength and type of influence that arousal has on performance.

This finding of Hodapp's (1982), in conjunction with the earlier findings, supports the theory that arousal related to emotions and to thoughts does affect performance in mathematics and science settings and that cognitions related to test anxiety also affect performance. This suggests that research concentrating solely on the cognitive or emotionality component of emotions may be overlooking part of the phenomenon that influences performance. It is apparent that both worry and emotionality have some effect on performance. What is less clear, perhaps, is how much influence each has. Another problem with examining emotionality and worry separately, even though they are thought by some to be conceptually independent, is that they covary in all studies that have included them, and attempts to manipulate one without influencing the other have not been very successful. Thus, it is apparent that worry, because it may occur prior to heightened arousal, does affect subsequent performance and predisposes one to experience emotionality, which, in turn, affects performance. To be comprehensive, classroom efforts to deal with strong emotional states that affect performance should address both the cognitive interference aspect and the increased arousal that accompany emotional states.

Treatment Strategies With Anxiety and Depression

The theories and experimental findings just described have helped researchers to develop ways to treat test anxiety, mathematics anxiety, and depression and to encourage students to improve their performance. Most of the studies described in this section concern test anxiety and mathematics anxiety and apply most specifically to classroom settings, although these treatment techniques may work in other settings as well. The treatments vary widely as to the dimensions of anxiety that they address. Some focus on emotional and physiological arousal, whereas others work to modify beliefs and expectancies that lead to poor performance (Dweck, 1975; Hembree, 1987). We first review approaches that focus on arousal, or on the emotionality component, and then discuss approaches directed at the worry or cognitive component and at how they compare with the affect-oriented procedures. We briefly mention studies concerning learned helplessness and attribution retraining that appear to address depression as well as anxiety. In each section, we evaluate the treatments in terms of their effectiveness at reducing anxiety and at enhancing test and school performance. Finally, we make suggestions as to how these techniques may be applied in a classroom setting.

Treatments Emphasizing the Emotionality Component

Hembree (1987) reviewed a number of experimental treatments for test anxiety. Treatments that focus on physiological arousal assume that arousal interferes with the performance of effortful tasks. Therefore, by reducing the amount of physiological arousal that a person is experiencing, a corresponding reduction of anxiety can be achieved. A highly aroused state preempts attentional and other processing resources that might otherwise be allocated to some cognitive task (Ellis & Ashbrook, 1988; Schaer & Isom, 1988). Hembree (1987) noted that the most frequently used emotionality-based treatment was systematic desensitization followed by various forms of relaxation training, and both reduced anxiety significantly. Spielberger and Vagg (1987) noted that, although systematic desensitization targets the physiological element of anxiety, it contains cognitive components as well. Subjects attend to relaxation cues and imagine being in the testing situation. Indeed, Spielberger and Vagg concluded that affect-based treatments by themselves were not effective in reducing anxiety. Systematic desensitization training had a greater effect than did relaxation training, suggesting that the cognitive component enhances its effectiveness.

The results of classroom test performance, in contrast, are somewhat mixed. Hembree (1987) found that systematic desensitization raised test scores as well as grade point averages. Relaxation training, although it raised test scores, did not improve overall grade point averages across studies. Schaer and Isom (1988) also obtained mixed results in their study of progressive relaxation. They tested the effect of this technique on both anxiety reduction and visual perception, and they assumed, as do the processing efficiency and resource allocation models described earlier (Ellis & Ashbrook, 1988; Eysenck & Calvo, 1992; Sarason, 1988), that humans have a limited processing capacity. Anxiety also interferes with the processing capacity of the visual system in much the same way as in more complex tasks. Schaer and Isom (1988) gave to hypnotized subjects suggestions that they would become less anxious and perform better in a disembedding task (the Group Embedded Figures Test [GEFT]). They found that posttest scores on the Test Anxiety Inventory decreased significantly but that scores on the GEFT were not significantly better than those of a control group receiving only suggestions related to test anxiety. These results are not surprising, given the arguments of Spielberger and Vagg (1987), who suggested that affect-based treatments may not be adequate in themselves in increasing test performance. Although test anxiety was alleviated in this case, it was not sufficient to raise performance.

One affect-based treatment that had positive effects on both anxiety and performance was examined by Smith and Nye (1989). This treatment, known as *induced affect*, contains a cognitive component. The subject imagines and tries to experience an emotion as strong as, or stronger than, that evoked by the testing situation, as well as suggestions of capability in controlling these feelings. Smith and Nye found this treatment to reduce anxiety significantly. In addition, lower state anxiety is related to better test performance, showing a facilitation effect through anxiety reduction for the induced affect procedure. It appears that affect-based procedures can be effective, particularly if they contain cognitive elements.

Cognitive-Based Treatments for Anxiety

The studies to be reviewed in this section emphasize the cognitive components of anxiety. One foundation of cognitive-based therapies is that certain kinds of thoughts interfere with one's attentional and performance capacities (Ellis & Ashbrook 1988; Hunsley, 1987; Kent & Jambunathan, 1989; Sarason, 1988). Self-referential thoughts, worries, and irrelevant thoughts use up some of the capacity for engaging in effortful tasks. Treatments addressing the worry component of anxiety aim to replace negative and unrealistic thoughts with more positive and realistic thoughts. Such is the goal of cognitive–behavioral therapy and rational–emotive therapy, two of the most common test anxiety treatments (Hembree, 1987; Spielberger & Vagg, 1987). These treatments can work both to free the mind of distracting thoughts and to encourage on-task behaviors and efforts that will lead to better test performance. Hembree (1987) found that cognitive–behavioral treatments were as effective as systematic desensitization and more effective than relaxation training in reducing test anxiety. These procedures affected both the worry and emotionality components of anxiety and actually enhanced *facilitating anxiety*, that is, the moderate amount of anxiety that can help performance of cognitively effortful tasks.

In addition, cognitively focused treatments enhance test performance to the same degree as affect-focused treatments (Hembree, 1987) and lead to increased grade point averages. One seemingly anomalous finding is that of Smith and Nye (1989), who found that a covert rehearsal procedure had a smaller effect on anxiety and performance than the more emotionality-focused, induced affect condition. One possible explanation for this finding is that covert rehearsal involves rehearsing the testing situation and the stress-reducing self-statements, but the emotions produced are less intense than they are in the induced affect condition. However, both techniques use encouraging cognitions and have proven moderately successful in reducing test anxiety. In fact, covert rehearsal reduced state anxiety in general to an even greater degree than did induced affect.

Some treatments, including the learned helplessness studies to be discussed next, may indirectly enhance performance by increasing feelings of self-efficacy. Smith (1989) exposed high test-anxious students to a multifaceted, coping skills training workshop conducted in a group setting. Cognitive restructuring (attempting to change a subject's self-schemata) and self-instructional statements formed the cognitive component of this treatment, which also included relaxation training. The results indicated that the treatment group scored reliably lower on the State-Trait Anxiety Inventory and higher in academic test performance than did a waiting-list control group. In addition, the treatment group exhibited an increase in scores on the Self-Efficacy Scale, suggesting that coping skills training positively influenced their beliefs concerning their ability to affect their outcomes. Coping skills training may work to increase one's sense of self-efficacy, thereby lowering anxiety and increasing perseverance, which leads to better performance. More generally, the studies reviewed by Hembree (1987) and Spielberger and Vagg (1987) reveal that replacing negative, self-defeating thoughts with positive, encouraging thoughts may prove fruitful. Hembree also noted that some students may need study-skills training in addition to anxiety-reducing techniques to enhance their performance.

Depression and Learned Helplessness

Another way to reduce test anxiety and enhance performance involves altering the students' attributional style. The studies reviewed here have implications for depression as well as for anxiety. According to Beck's (1967) cognitive theory of depression, negative thoughts and negative self-schemata may cause anxiety or depressed feelings, leading to poor task performance and to the validation of that negative self-schema. Learned helplessness results when the student gives up expending effort to succeed at such tasks (Fincham, Hokoda, & Sanders, 1989). We briefly discuss studies that have attempted and largely succeeded in altering students' attributions to failure and, consequently, improved their classroom performance.

Dweck (1975) conducted one of the first such studies in an attempt to enhance elementary school children's performance on mathemathics problems. She noticed that some subjects tended to attribute performance outcomes to lack of ability or to external, uncontrollable factors (e.g., luck). Following failure, these subjects often did not attempt to solve easy tasks, a behavior that often accompanies depression. She divided these subjects into two treatment groups, one that received very easy mathematics problems (success only) and one that received some failure trials (attribution retraining). For both groups, failure trials were inserted at the middle and at the end of the 25-day training period. For the attribution retraining group, the experimenter told the subjects (upon a scheduled failure) that they should have tried harder, whereas the success-only group received no such feedback. Dweck found that the success-only group showed performance decrements following failure, whereas those in the attribution retraining group showed negligible impairment or even improvement following failure trials. They also began to attribute their outcomes to effort more than the success-only group did. It appears, then, that the attribution of effort is a successful coping mechanism that encourages greater perseverance and enhanced performance.

Fowler and Peterson (1981) replicated and extended Dweck's (1975) findings in a study containing two treatment groups. The first group received feedback from the experimenter attributing their performance to effort. The second group listened to a tape recording of a male or female voice saying to themselves, "I got that one right. That means I tried hard," or, following failure, "I didn't get that one right. That means I have to try harder." The subjects then read technical sentences, some of which were beyond their ability. Subjects in the second group were told that the aforementioned attributions were good things to say to themselves and were told to do so after each scheduled success or failure. Both groups showed increased persistence on the tasks and increased tendencies to attribute outcomes to effort than did a no-feedback control group. In general, children seem to be able to internalize from a teacher as well as from a peer role model and change their attributional style accordingly.

Craske (1985) used a similar procedure in an observational learning study. Children who were identified as helpless watched a videotape of a same-aged child performing a puzzle task and received effort-attributional feedback similar to that in the Fowler and Peterson (1981) study just described. Subjects exposed to this treatment exhibited greater persistence in attempting to solve puzzles than did the control group; this effect was also more pronounced for girls than for boys. Attribution retraining, then, may be effective in modifying one's self-schemata so that negative expectations can be replaced with positive ones that will encourage greater effort toward succeeding at a given task. These treatments may also alleviate the interfering effects of worry-related cognitions and free-up more processing capacity for the task, which would be expected on the basis of the assumptions of the resource allocation model (Ellis & Ashbrook, 1988). Therefore, telling oneself to try harder may keep one on task and minimize interfering thoughts; changing one's thoughts can lead to an improved mood and improved performance. Worry cognitions may be a cause of physiological arousal, leading to increased salience of a mood, reduced allocation of processing resources, and increased cognitive interference. Accordingly, eliminating these worry cognitions reduces this interference and enhances processing capacity and task performance.

Applications in School Settings

Most of the techniques mentioned earlier may be applied in classroom settings, and we briefly discuss several of them. Progressive deep muscle relaxation is a common element in most stress-management programs (Rosen, 1977) and involves tensing and relaxing particular groups of muscles in sequence. High test-anxious students may find this technique useful before a test, which, at the very least, can reduce anxiety and may facilitate performance. Meditation, although not discussed earlier, may also help eliminate distracting thoughts and help one to confront and change self-defeating thoughts. Attribution retraining can become part of the classroom curriculum along with study-skills training. Teachers can administer difficult problems in mathematics or other subject areas and teach students to say to themselves some of the statements used in the aforementioned studies (Craske, 1985; Fowler & Peterson, 1981) to reattribute their outcomes to effort and boost performance. Finally, multifaceted treatments, such as that used in van der Ploeg-Stapert and van der Ploeg (1987), can become more available in schools. Because these programs have been effective in experimental studies, it may be profitable to apply them on a wider scale in school settings. In general, the vast array of cognitive–behavioral interventions developed in clinical and other settings are potentially applicable, in some adapted form, to classroom settings.

So far, our focus has been on the usefulness of principles, theories, and interventions derived from research on cognition and emotion. Procedures and strategies designed to reduce worry and undue arousal certainly have their value in classroom settings. Equally important, however, are approaches that focus on the building of competencies, skills, and knowledge. There is a good deal of evidence that increasing the competence of students produces the additional positive by-product of reducing worry, anxiety, and other aversive emotional states. Thus, we wish to emphasize the importance of building competencies in students not only as obviously desirable in its own right but also as having additional desirable effects. Finally, it is important to comment on the role of anxiety and negative emotional states in teachers, whose affective states clearly affect students. Certainly, teachers are sometimes anxious and thus may fail to encourage discussion, independent thinking, or other important behaviors. Fortunately, there are ways to teach so that anxiety, worry, or negative moods states in students can be minimized. These points, of course, emphasize the pervasive role that *context* plays in all teaching and learning.

Conclusion

This chapter has described some of the principal theories of cognition and emotion, reviewed some of the classroom learning literature, and described some of the strategies derived from this research and theory that are applicable to classroom settings. The focus of this chapter was on ways to reduce worry and emotionality in the classroom so as to facilitate learning. A variety of approaches have been adopted, and varying degrees of success have been achieved by cognitive–behavioral procedures. Finally, the importance of building students' competence and alleviating teacher anxiety were briefly noted as important issues.

References

Beck, A. T. (1967). *Depression: Clinical, experimental, and theoretical aspects*. New York: Harper & Row.

Beck, A. T., Rush, A. J., Shaw, B. S., & Emery, G. (1979). *Cognitive therapy of depression*. New York: Guilford Press.

Blaney, P. H. (1986). Affect and memory: A review. *Psychological Bulletin, 29*, 229–248.

Blankstein, K. R., Flett, G. L., Boase, P., & Toner, B. B. (1990). Thought listing and endorsement measures of self-referential thinking in test anxiety. *Anxiety Research, 2*, 103–111.

Bower, G. H. (1981). Mood and memory. *American Psychologist, 36*, 129–148.

Bower, G. H., & Mayer, J. D. (1989). In search of mood-dependent memory [Special issue]. *Journal of Social Behavior and Personality, 4*(2), 121–156.

Carver, C. C., Scheier, M. F., & Klahr, D. (1987). Further explanations of a control-process model of test anxiety. In R. Schwarzer, H. M. van der Ploeg, & C. D. Spielberger (Eds.), *Advances in test anxiety research* (Vol. 5). Lisse, The Netherlands: Swets & Zeitlinger.

Clark, M. S., Milberg, S., & Erber, R. (1988). Arousal-state-dependent memory: Evidence and implications for understanding social judgments and social behavior. In K. Fiedler & J. Forgas (Eds.), *Affect, cognition, and social behavior* (pp. 63–83). Göttingen, Federal Republic of Germany: Hogrefe.

Craske, M. L. (1985). Improving persistence through observational learning and attribution retraining. *British Journal of Educational Psychology, 55*, 138–147.

Deffenbacher, J. L. (1978). Worry, emotionality, and task generated interference: An empirical test of attentional theory. *Journal of Educational Psychology, 70*, 248–254.

Dweck, C. S. (1975). The role of expectations and attributions in the alleviation of learned helplessness. *Journal of Personality and Social Psychology, 31*, 674–685.

Easterbrook, J. A. (1959). The effect of emotion on cue utilization and the organization of behavior. *Psychological Review, 66*, 183–201.

Eich, E. (1989). Theoretical issues in state-dependent memory. In H. L. Roediger & F. I. M. Craik (Eds.), *Varieties of memory and consciousness* (pp. 000–000). Hillsdale, NJ: Erlbaum.

Ellis, H. C. (1985). On the importance of mood intensity and encoding demands in memory: Commentary on Hasher, Rose, Zacks, Sanft, and Doren. *Journal of Experimental Psychology: General, 114*, 392–395.

Ellis, H. C. (1990). Depressive deficits in memory: Processing initiative and resource allocation. *Journal of Experimental Psychology: General, 119*, 60–62.

Ellis, H. C. (1991). Focused attention and depressive deficits in memory. *Journal of Experimental Psychology: General, 119*, 310–312.

Ellis, H. C., & Ashbrook, P. W. (1988). Resource allocation model of the effects of depressed mood states on memory. In K. Fiedler & J. F. Forgas (Eds.), *Affects, cognition and social behavior* (pp. 25–43). Göttingen, Federal Republic of Germany: Hogrefe.

Ellis, H. C., & Ashbrook, P.W. (1989). The "state" of mood and memory research: A selective review [Special issue]. *Journal of Social Behavior and Personality, 4*(2), 1–21.

Ellis, H. C., & Franklin, J. B. (1983). Memory and personality: External versus internal locus of control and superficial organization in free recall. *Journal of Verbal Learning and Verbal Behavior, 22*, 61–74.

Ellis, H. C., & Hertel, P. T. (1993). Cognition, emotion and memory: Some applications and issues. In C. Izawa (Ed.), *Cognitive psychology applied* (pp. 000–000). Hillsdale, NJ: Erlbaum.

Ellis, H. C., & Hunt, R. R. (1993). *Fundamentals of cognitive psychology* (5th ed.). Madison, WI: Brown & Benchmark.

Ellis, H. C., Seibert, P. S., & Herbert, B. J. (1990). Mood state effects on thought listing. *Bulletin of the Psychonomic Society, 28*, 147–150.

Ellis, H. C., Thomas, R. L., McFarland, A. D., & Lane, J. W. (1985). Emotional mood states and retrieval in episodic memory. *Journal of Experimental Psycholoy: Learning, Memory, and Cognition, 11*, 363–370.

Ellis, H. C., Thomas, R. L., & Rodriguez, I. A. (1984). Emotional mood states and memory: Elaborative encoding, semantic processing, and cognitive effort. *Journal of Experimental Psychology: Learning, Memory, and Cognition, 10*, 470–482.

Eysenck, M. W. (1982). *Attention and arousal cognition and performance.* Berlin: Springer-Verlag.

Eysenck, M. W., & Calvo, M. G. (1992). Anxiety and performance: The processing efficiency theory. *Cognition and Emotion, 6,* 409–434.

Fiedler, K., & Forgas, J. (Eds.). (1988). *Affect, cognition, and social behavior.* Göttingen, Federal Republic of Germany: Hogrefe.

Fincham, F. D., Hokoda, A., & Sanders, R. (1989). Learned helplessness, test anxiety, and academic achievement: A longitudinal analysis. *Child Development, 60,* 138–145.

Fowler, J. W., & Peterson, P. I. (1981). Increasing reading persistence and altering attributional style of learned helplessness children. *Journal of Educational Psychology, 73,* 251–260.

Geen, R. G. (1987). Test anxiety and behavioral avoidance. *Journal of Research in Personality, 21,* 481–488.

Hasher, L., & Zacks, R. T. (1988). Working memory, comprehension, and aging: A review and a new view. In G. H. Bower (Ed.), *The psychology of learning and motivation* (pp. 193–225). San Diego, CA: Academic Press.

Hembree, R. (1987). Correlates, causes, effects, and treatment of test anxiety. *Review of Educational Research, 58,* 47–77.

Herman, W. E. (1990). Fear of failure as a distinctive personality trait measure of test anxiety. *Journal of Research and Development in Education, 23*(3), 180–185.

Hertel, P. T., & Hardin, T. S. (1990). Remembering with and without awareness in a depressed mood: Evidence of deficits in initiative. *Journal of Experimental Psychology: General, 119,* 45–59.

Hertel, P. T., & Rude, S. S. (1991). Depressive deficits in memory: Focusing attention improves subsequent recall. *Journal of Experimental Psychology: General, 120,* 301–309.

Hodapp, V. (1982). Causal interference from non-experimental research on anxiety and educational achievement. In H. W. Krohne & L. Laux (Eds.), *Achievement, stress, and anxiety.* Washington, DC: Hemisphere.

Humphreys, M. S., & Revelle, W. (1984). Personality, motivation, and performance: A theory of the relationship between individual differences and information processing. *Psychological Review, 91,* 3–184.

Hunsley, J. (1987). Cognitive processes in mathematics anxiety and test anxiety: The role of appraisals, internal dialogue, and attributions. *Journal of Educational Psychology, 79,* 388–397.

Ingram, R. E. (1984). Toward an information processing analysis of depression. *Cognitive Therapy and Research, 8,* 443–478.

Jacoby, L. L. (1978). On interpreting the effects of repetition: Solving a problem versus remembering a solution. *Journal of Verbal Learning and Verbal Behavior, 17,* 649–667.

Kahneman, D. (1973). *Attention and effort.* Englewood Cliffs, NJ: Prentice Hall.

Kent, G., & Jambunathan, P. (1989). A longitudinal study of the intrusiveness of cognitions in test anxiety. *Behavior Research and Therapy, 27*(1), 43–50.

Kuiken, D. (Ed.). (1989). Mood and memory: Theory, research, and applications [Special issue]. *Journal of Social Behavior and Personality, 4*(2).

Kuiken, D. (Ed.). (1991). *Mood and memory.* Newbury Park, CA: Sage.

Kuiper, N. A., MacDonald, M. R., & Derry, P. A. (1983). Parameters of a depressive self-schema. In J. Suls & A. G. Greenwald (Eds.), *Psychological perspectives on the self* (Vol. 2, pp. 191–217). Hillsdale, NJ: Erlbaum.

Liebert, R. M., & Morris, L. W. (1967). Cognitive and emotional components of test anxiety. *Psychological Reports, 20,* 975–978.

Mandler, G., & Sarason, I. G. (1952). A study of anxiety and learning. *Journal of Abnormal and Social Psychology, 47,* 166–173.

McFarland, C. E., Jr., Frey, T. J., & Rhodes, D. D. (1980). Retrieval of internally versus externally generated words in episodic memory. *Journal of Verbal Learning and Verbal Behavior, 19,* 210–225.

Morris, L. W., & Liebert, R. M. (1969). The effects of anxiety on timed and untimed intelligence tests: Another look. *Journal of Consulting and Clinical Psychology, 33,* 240–244.

Morris, L. W., & Liebert, R. M. (1970). Relationship of cognitive and emotional components of test anxiety to physiological arousal and academic performance. *Journal of Consulting and Clinical Psychology, 35,* 332–337.

Neumann, O. (1987). Beyond capacity: A functional view of attention. In H. Heuer & A. F. Sanders (Eds.), *Perspectives on perception and action* (pp. 361–394). Hillsdale, NJ: Erlbaum.

O'Brien, E. J., & Myers, J. L. (1985). When comprehension difficulty improves memory for text. *Journal of Experimental Psychology: Learning, Memory, and Cognition, 11,* 12–21.

Ramirez, O. M., & Dockweiler, C. J. (1985). Mathematics anxiety: A systematic review. In R. Schwarzer, H. M. van der Ploeg, & C. D. Spielberger (Eds.), *Advances in test anxiety research* (Vol. 5). Lisse, The Netherlands: Swets & Zeitlinger.

Revelle, W., & Loftus, D. A. (1990). Individual differences and arousal: Implications for the study of mood and memory. *Cognition and Emotion, 4,* 209–237.

Rosen, G. (1977). *The relation book.* Englewood Cliffs, NJ: Prentice Hall.

Sarason, I. G. (1978) The Test Anxiety Scale: Concept and research. In C. D. Spielberger & I. G. Sarason (Eds.), *Stress and anxiety* (Vol. 5, pp. 193–216). Washington, DC: Hemisphere.

Sarason, I. G. (1984). Stress, anxiety, and cognitive interference: Reactions to tests. *Journal of Personality and Social Psychology, 46,* 929–938.

Sarason, I. G. (1988). Anxiety, self-preoccupation, and attention. *Anxiety Research, 1,* 3–7.

Sarason, I. G., & Stoops, R. (1978). Test anxiety and the passage of time. *Journal of Consulting and Clinical Psychology, 46,* 102–109.

Schaer, B., & Isom, S. (1988). Effectiveness of progressive relaxation on test anxiety and visual perception. *Psychological Reports, 63*, 511–518.

Schwarzer, R., van der Ploeg, H. M., & Spielberger, C. D. (Eds.). (1987). *Advances in test anxiety research* (Vol. 5). Lisse, The Netherlands: Swets & Zeitlinger.

Seibert, P. S., & Ellis, H. C. (1991). Irrelevant thoughts, emotional mood states and cognitive task performances. *Memory & Cognition, 19*, 507–513.

Sewell, T. E., Farley, F. H., & Sewell, F. B. (1983). Anxiety, cognitive style, and mathematics achievement. *Journal of General Psychology, 109*, 59–66.

Smith, R. E. (1989). Effects of coping skills training on generalized self-efficacy and locus of control. *Journal of Personality and Social Psychology, 56*, 228–233.

Smith, R. E., & Nye, S. L. (1989). Comparison of induced affect and covert rehearsal in the acquisition of a stress management coping skills. *Journal of Counseling Psychology, 36*, 17–33.

Spielberger, C. D., Anton, W. D., & Bedell, J. (1976). The nature and treatment of test anxiety. In M. Zukerman & C. D. Spielberger (Eds.), *Emotions and anxiety: New concepts, methods and applications*. Hillsdale, NJ: Erlbaum.

Spielberger, C. D., Gorsuch, R., & Lushene, R. (1970). *Manual for the State-Trait Anxiety Inventory*. Palo Alto, CA: Consulting Psychologists Press.

Spielberger, C. D., & Vagg, P. R. (1987). The treatment of test anxiety: A transactional process model. In R. Schwarzer, H. M. van der Ploeg, & C. D. Spielberger (Eds.), *Test anxiety research* (Vol. 5). Lisse, The Netherlands: Swets & Zeitlinger.

Swanson, H. L. (1984). Effects of cognitive effort and word distinctiveness on learning disabled and nondisabled readers' recall. *Journal of Educational Psychology, 76*, 894–908.

Teasdale, J. D., & Fogarty, S. J. (1979). Differential effects of induced mood on retrieval of pleasant and unpleasant events from episodic memory. *Journal of Abnormal Psychology, 88*, 248–257.

Teasdale, J. D., & Russell, M. L. (1983). Differential effects of induced mood on the recall of positive, negative and neutral words. *British Journal of Clinical Psychology, 22*, 163–172.

Tyler, S. W., Hertel, P. T., McCallum, M. C., & Ellis, H. C. (1979). Cognitive effort and memory. *Journal of Experimental Psychology: Human Learning and Memory, 5*, 607–617.

van der Ploeg-Stapert, J. D., & van der Ploeg, H. M. (1987). The evaluation and follow-up of a behavioral group treatment of test anxious adolescents. In R. Schwarzer, H. M. van der Ploeg, & C. D. Spielberger (Eds.), *Test anxiety research* (Vol. 5). Lisse, The Netherlands: Swets & Zeitlinger.

Williams, J. M. G., Watts, F. N., MacLeod, C., & Mathews, A. (1988). *Cognitive psychology and emotional disorders*. New York: Wiley.

Witkin, H., Oltman, P., Raskin, E., & Karp, S. (1971). *Manual for the Embedded Figures Test*. Palo Alto, CA: Consulting Psychologists Press.

CHAPTER 4

The Jasper Series: Theoretical Foundations and Data on Problem Solving and Transfer

Cognition and Technology Group at Vanderbilt[1]

Cognition and Technology Group at Vanderbilt

During the past several years, the Cognition and Technology Group at Vanderbilt (CTGV) has had the opportunity to develop and to test educational programs that are based on psychological principles of learning and that are designed to motivate students to learn to think for themselves and to solve realistic, complex problems. We focus on the goals of improving thinking and problem solving because there is considerable evidence that today's students are not particularly strong in these areas (e.g., Bransford, Goldman, & Vye, 1991; Nickerson, 1988; Resnick, 1987). We use technology (especially computer, videodisc, and teleconferencing

[1]Members of the Cognition and Technology Group who contributed to this chapter are John D. Bransford, Susan R. Goldman, Ted S. Hasselbring, James W. Pellegrino, Susan M. Williams, and Nancy Vye.

Preparation of this chapter was supported, in part, by grants from the James S. McDonnell Foundation (No. 87-39) and the National Science Foundation (No. NSF-MDR 9050191 and No. NSF-MDR-9252990). The ideas expressed herein do not necessarily reflect views of these agencies.

technologies) because different learning goals require different types of learning activities (e.g., Jenkins, 1979) and because new technologies have the potential to support teaching and learning activities that help students learn to think and to solve problems. Our goal in this chapter is to provide an overview of one of our programs, the theoretical framework that guides its ongoing development, and some of the data collected that provide information about human learning and problem solving.

The chapter is divided into four sections: (a) discussion of the theoretical framework that motivates our work; (b) description of a particular example of our theoretical framework—the Jasper Woodbury problem-solving series; (c) an overview of the program of research that accompanies the Jasper Series plus discussion of key findings from one aspect of this research program that focuses on issues of learning and transfer; and (d) discussion of ways that our research on transfer has led us to refine our materials.[2]

Theoretical Foundations

The theoretical framework that guides our work is consistent with a class of theories called *constructivist theories* (e.g., Bransford & Vye, 1989; Clement, 1982; Cobb, Yackel, & Wood, 1992; Duffy & Bednar, 1991; Minstrell, 1989; Perkins, 1991; Resnick & Klopfer, 1989; Scardamalia & Bereiter, 1991; Schoenfeld, 1989; Spiro, Feltovich, Jacobson, & Coulson, 1991). Theorists who emphasize the constructive nature of learning argue for the need to change the nature of the teaching and learning processes that occur in most classrooms. Instead of having teachers "transmit" information that students "receive," these theorists emphasize the importance of having students become actively involved in the construction of knowledge. For example, constructivist theorists want to help students to learn to construct and to coordinate effective problem representations through the use of symbolic and physical models, through reasoning and argu-

---

[2]Discussion in the first two sections borrows heavily from previous descriptions found in CTGV (1992b, 1992c, 1993); subsequent sections are unique to this chapter. Because of page constraints, we are unable to provide a full description of our theoretical framework, program components, and research findings. We indicate these omissions as we come to them and refer readers to relevant papers for further information.

mentation, and through deliberate application of problem-solving strategies (e.g., Brown, Collins, & Duguid, 1989; Clement, 1982; Minstrell, 1989;
Palincsar & Brown, 1989; Resnick & Klopfer, 1989; Scardamalia & Bereiter,
1991; Schoenfeld, 1989). A basic assumption of the constructivist position
is that students cannot learn to engage in effective knowledge-construc-
tion activities simply by being told new information (e.g., Bransford,
Franks, Vye, & Sherwood, 1989). Instead, students need repeated opportunities to engage in in-depth exploration, assessment, and revision of
their ideas over extended periods of time. We discuss several aspects of
our version of constructive learning theory.

Generative Learning

A number of theorists emphasize the importance of helping students
engage in *generative* rather than *passive* learning activities. In physics,
for example, many students bring a number of preconceptions to learning,
some of which are misconceptions (e.g., Clement, 1982; Minstrell, 1989).
For students to overcome these misconceptions, it is not sufficient simply
to memorize how a scientist represents and explains various phenomena.
Rather, students need to engage in argumentation and reflection as they
try to use and then to refine their existing knowledge so as to make sense
of alternate points of view.

In the mathematics domain, efforts to reform the nature of mathematics curricula also stress that mathematics classrooms need to shift
from an emphasis on the teacher imparting knowledge to one in which
students attempt to use their current skills and knowledge to approach
problems to be solved (e.g., Charles & Silver, 1988; National Council of
Teachers of Mathematics [NCTM], 1989; Schoenfeld, 1985, 1989; Yackel,
Cobb, Wood, Wheatley, & Merkel, 1990). Even the ability to learn effectively from worked-out examples involves the construction of links to
other aspects of one's knowledge (e.g., Chi, Bassok, Lewis, & Glaser, 1989).
As Resnick and Resnick (1991) noted, for concepts and principles to be
learned effectively, "they must be used generatively—that is, they have
to be called upon over and over again as ways to link, interpret and
explain new information" (p. 41). Silver (1990) discussed several types
of materials that might encourage such activities, including worked ex-

amples that involve active construction, open-ended problems, and problems in which concepts and procedures are connected.

Findings from a number of studies suggest that knowledge that is not acquired and used generatively tends to become what Whitehead (1929) called "inert knowledge"—knowledge that is not used spontaneously even though it is relevant (e.g., Bransford et al., 1989; Gick & Holyoak, 1980, 1983; Scardamalia & Bereiter, 1985). Data indicate that knowledge is less likely to remain inert when it is acquired in a problem-solving mode rather than in a factual-knowledge mode (Adams et al., 1988; Lockhart, Lamon, & Gick, 1988).

Anchored Instruction

The approach to instruction that we have been exploring is called *anchored instruction* (CTGV, 1990, 1991). The essence of the approach is to "anchor" or to "situate" instruction in the context of meaningful problem-solving environments that allow teachers to simulate in the classroom some of the advantages of "in-context" apprenticeship training (Brown et al., 1989). We refer to our anchored environments as *macrocontexts* because they involve complex situations that require students to formulate and to solve a set of interconnected subproblems (Bransford, Sherwood, & Hasselbring, 1988). In addition, each of the anchors can be viewed from multiple perspectives. In contrast, problem sets, such as the ones that occur at the end of chapters in textbooks, typically involve a series of disconnected microcontexts, one for each problem in the problem set. Our anchored instruction approach shares a strong family resemblance to many instructional programs that are case based and problem based (e.g., Barrows, 1985; Gragg, 1940; Spiro et al., 1991; Williams, 1991).

A primary goal that guides our selection and development of anchors is to allow students who are relative novices in an area to experience some of the advantages available to experts when they are trying to learn new information about their area (e.g., Bransford, Sherwood, Hasselbring, Kinzer, & Williams, 1990; CTGV, 1990). Theorists such as Dewey (1933),

Schwab (1960), and Hanson (1970) emphasize that experts in an area have been immersed in phenomena and are familiar with how they have been thinking about them. When introduced to new theories, concepts, and principles that are relevant to their areas of interest, the experts can experience the changes in their own thinking that these ideas afford. For novices, however, the introduction of concepts and theories often seems like the mere introduction of new facts or of mechanical procedures to be memorized. Our approach is to have novices use their available knowledge first to attempt to understand the phenomena and activities depicted in an anchor and then to be able to experience the changes in their own noticing and understanding as they are introduced to concepts and to theories that are relevant to the anchors. We especially emphasize the importance of helping novices move from a general understanding of a complex problem to one in which they learn to generate and define the distinct subgoals necessary to achieve an overall goal.

Cooperative Learning and Generativity

Our anchors are also designed with an eye toward their use in cooperative learning settings. A number of theorists argue that cooperative learning and cooperative problem-solving groups enhance opportunities for generative learning. In cooperative groups, students have the opportunity to form communities of inquiry that allow them to discuss and explain and, hence, learn with understanding (e.g., NCTM, 1989; Palincsar & Brown, 1984, 1989; Vygotsky, 1978).

We are also aware of potential disadvantages of cooperative learning groups. For example, Salomon and Globerson (1989) discussed many pitfalls in the implementation of groups in classrooms. In addition, some groups work more effectively than others and members experience differential benefits of group work (Cosden, Goldman, & Hine, 1990; Goldman, Cosden, & Hine, 1987, 1992; Hine, Goldman, & Cosden, 1990). One of our ongoing research goals is to understand better how to help teachers to provide scaffolds (degrees of structure) that enable initially less skilled groups to begin to explore ideas without going too far astray, yet to help them become generative learners who are self-directed rather than teacher-directed.

The Jasper Series as an Example of Anchored Instruction

As noted earlier, the CTGV is experimenting with a variety of anchored instruction programs that focus on specific content areas while also providing opportunities for cross-curricular extensions (e.g., CTGV, 1993, in press). We have created anchors, plus instructional activities to accompany them, that focus on the areas of mathematics (e.g., CTGV, 1991, 1992a, 1992b), science (e.g., CTGV, 1993; Hickey, Pellegrino, Petrosino, & the CTGV, 1991), and literacy (e.g., CTGV, in press; McLarty et al., 1990; Sharp et al., 1992). In this chapter, we emphasize our Jasper Woodbury Series, which focuses on mathematical problem solving, with extensions to science, history, geography, and other subjects (CTGV, 1993). We choose to discuss the Jasper series because, to date, it has been more widely used than our other programs and has generated the most research.

The *Adventures of Jasper Woodbury* is a video-based series designed to promote problem posing, problem solving, reasoning, and effective communication. Each adventure is a 15–20-minute story. At the end of each story, the major character (or group of characters) is faced with a challenge that the students in the classroom must solve before they are allowed to see how the movie characters solved the challenge. The adventures are designed in pairs, with each pair sharing a common problem schema. The first pair of adventures deals with issues of trip planning; the second pair involves generating a business plan using statistics; and the third involves meaningful uses of geometry. We have also created extension videos to accompany each adventure. (Examples of extensions are discussed in more detail later.)

One link between our theoretical framework and our materials development is Gibson's (1977) concept of *affordances*. Gibson argued that different features of the environment afford classes of activities for particular organisms such as "walk-onable," "climbable," "swim-able," and so forth. Our notion is that instructional materials can be viewed from a similar perspective. Different types of materials are differentially effective for helping students engage in particular kinds of learning activities (Jenkins, 1979). Some problems lend themselves to protracted problem posing and formulation, whereas other problems structure the situation completely, leaving little to do but "add up the numbers."

Elsewhere (CTGV, 1992a), we note that the primary method of teaching problem solving in mathematics involves word problems and that these have limitations. Traditional word problems often provide the goal and only those numbers needed to solve the problem; hence, they afford only computational selection. In other cases, even the type of computation is made clear to the students (because the chapter focuses on a particular operation, such as addition), so the word problems actually provide only computational practice (Porter, 1989). Our Jasper adventures afford activities such as generating subgoals, identifying relevant information, cooperating with others to plan and to solve complex problems, discussing the advantages and disadvantages of possible solutions, and comparing perspectives by pointing out and explaining interesting events. Of course, the mere existence of these affordances does not guarantee that afforded activities will occur. The degree to which these affordances are realized depends on the teaching model that one adopts in the context of Jasper. (Space does not permit a description of the teaching and learning activities that accompany Jasper in the classroom. Discussions of different teaching models appear in CTGV, 1992b.)

The types of learning activities that we want our materials to support are consistent with recommendations suggested by the NCTM's Commission on Standards for School Mathematics. The NCTM's (1989) suggestions for changes in classroom activities include more emphasis on complex, open-ended problem solving, communication, and reasoning; more connections from mathematics to other subjects and to the world outside the classroom; and more uses of calculators and powerful computer-based tools, such as spreadsheets and graphing programs for exploring relationships (as opposed to having students spend an inordinate amount of time calculating by hand). In proposing a more generative approach to mathematics learning, the NTCM stated the following:

> [The] mathematics curriculum should engage students in some problems that demand extended effort to solve. Some might be group projects that

require students to use available technology and to engage in cooperative problem solving and discussion. For grades 5–8 an important criterion of problems is that they be interesting to students. (p. 75)

Design Principles for the Jasper Series

Our attempt to create instructional materials that afford generative learning activities has been guided by the following seven basic design principles: (a) video-based format; (b) narrative with realistic problems (rather than a lecture on video); (c) generative format (i.e., students must generate the subproblems to be solved at the end of each story); (d) embedded data design (i.e., all the data needed to solve the problems are in the video); (e) problem complexity (i.e., each adventure involves a problem of at least 14 steps); (f) pairs of related adventures (to discuss issues of transfer); and (g) links across the curriculum. These design principles are discussed in more detail elsewhere (CTGV, 1991, 1992a; Hickey et al., 1991; McLarty et al., 1990). For present purposes, it is sufficient to note that the design principles mutually influence one another and operate as a gestalt rather than as a set of independent features of the materials. For example, the narrative format, the generative design of the stories, and the fact that the adventures include embedded data make it possible for students to learn to generate subgoals, find relevant information, and engage in reasoned decision making. The complexity of the problems helps students deal with this important aspect of problem solving, and the use of video helps make the complexity manageable. The video format also makes it easier to embed the kinds of information that provide opportunities for links across the curriculum. The video is also important because it brings the world into the classroom in a manner that motivates students, and it makes complex mathematical problem solving accessible to students who have difficulties imagining complex situations by reading. The pairs of related adventures afford discussions about transfer.

The Research Program Associated With the Jasper Series

Although the Jasper Series has become a product that can be used in schools, one of its most important functions is to provide opportunities for laboratory and school-based research on issues of learning, instruction, assessment, and transfer. There are three major components of our Jasper research program. The first involves theory-based materials development, where the materials include the adventures and teaching materials to accompany them. Earlier in this chapter, we discussed the general constructivist framework that guides the development of our materials, and we summarized the set of design principles that afford (cf. Gibson, 1977) new teaching and learning activities in the classroom. As we discuss later, much of our research is designed to inform our theory of learning and transfer and our materials development so that we may continue to improve our abilities to help people learn to learn.

The second major component of our research on Jasper involves an analysis of issues of broad-scale assessment. Most people agree that we cannot hope to make teaching and learning in schools more generative and thoughtful without changing the nature of the assessments that school systems use for formative and summative evaluation and accountability (e.g., Bransford et al., 1991; Goldman, Pellegrino, & Bransford, in press; Resnick & Resnick, 1991; Wolf, Bixby, Glen, & Gardner, 1991). In particular, assessments that focus only on discrete skills and knowledge structures provide no information about the degree to which students can coordinate their knowledge to accomplish realistic, complex goals.

Some of our initial work on assessing the effects of the Jasper Series has been conducted in the context of a broad-scale implementation project, in which Jasper was used in schools in nine different states. The assessments focused on changes in students' abilities to solve pencil-and-paper problems, plus changes in their attitudes toward complex problem solving and mathematics. The results of our assessment data have been very positive and are reported in CTGV (1992b) and in Pellegrino et al., (1991). Nevertheless, these assessments were essentially summative rather than formative in nature; they were not used throughout the school year to help teachers and students to assess continually their current levels of progress and to modify learning and instructional activities as needed. One of our current projects on assessment is to focus on ways of using distance-learning technologies (e.g., satellites, two-way videoconferencing) to create interesting formative assessment frameworks that will work on a broad scale (e.g., CTGV, 1992d).

The third major component of our work on Jasper involves cognitive analyses (generally through the analysis of protocol data) of changes in people's abilities to represent and to solve complex problems as they gain experience with the Jasper Series. We are particularly interested in exploring relationships between people's abilities to transfer to new problems as a function of the nature of the teaching and learning activities that take place in the classroom. This line of research includes studies of the effects on transfer of working with Jasper versus more traditional word problems, plus the effects of different approaches to instruction that are used in the context of Jasper (see CTGV [1992b] for a discussion of different approaches to instruction in the Jasper context). Because it is important to understand how teachers and students understand Jasper problems, our research on cognitive processes includes studies of adults.

Overview of Specific Studies and Issues

In the following discussion, we summarize findings from the third component of our Jasper research project—the one that focuses on cognitive processes of learning, instruction, and transfer. Because of space constraints, our discussion includes only a subset of our studies in this area; for example, we focus only on the first two Jasper adventures and do not attempt to discuss studies of the effects of group compared with individual attempts to solve Jasper problems (e.g., Barron, 1991; Lamon, 1992; Rewey & the CTGV, 1991). Detailed analyses of the data from the studies that we present are available in Goldman et al. (1992) and Williams, Bransford, Vye, Goldman, and Carlson (1992). The issues to be considered and our expectations about what we would find are as follows:

1. To what extent are our schools preparing students to solve complex, real-world problems like those found in the Jasper Series? Our expectations, which unfortunately are strongly confirmed by our data, were that students are underprepared for problems such as these.
2. To what extent is the ability to transfer to new, complex problems influenced by instruction that focuses on complex problems such as those in Jasper compared with instruction that involves the same con-

cepts as Jasper (e.g., distance, rate, and time), yet is organized around the typical one- and two-step word problems found in most mathematics curricula? Our expectations were consistent with the principle of *transfer appropriate processing* (e.g., Bransford, Franks, Morris, & Stein, 1979) and led us to predict that the best way to learn to deal with complexity is, in fact, to deal with complexity. Because of the visual format of the Jasper problems, we also assumed that the complexity would be manageable by students in the fifth and sixth grades. Moreover, we hoped that students who worked in the context of Jasper would also be able to solve word problems on distance–rate–time even though they had not practiced on these.

3. To what extent is transfer from one Jasper adventure to another based on general principles of problem solving versus more specific similarities in problem-solving procedures? Our expectations were that some general transfer effects would be evident (e.g., a willingness to persist in the face of complexity) but that specificity of transfer would be the rule.

4. To what extent does the ability to solve a Jasper problem and transfer to an analogous problem guarantee that students have developed a deep understanding of the problem spaces that characterize the Jasper and the transfer problems? As our work progressed, we began to suspect that the ability to transfer to analogous problems did not guarantee deep understanding of the original Jasper problem, and our data confirm our suspicions. Because of these findings, we have modified our Jasper materials and describe their current state.

Planning Net Analyses of the Jasper Problems

Investigations of the issues discussed earlier require detailed analyses of students' and teachers' mental representations of the solutions to different Jasper problems. Jasper problems are complex and require extensive planning to solve. Multiple solution plans must be generated, and to decide among different plans, students need to assess the feasibility of each plan by testing whether it meets constraints inherent in the problem. Furthermore, because most of the Jasper problems are designed so that multiple plans are feasible, it is necessary for students to select among plans by considering each relative to a set of optimization criteria. Our approach to describing how people mentally represent these complex problems has been first to construct an "ideal" representation describing the plans, goals, and decisions that would need to be considered to solve each Jasper episode and then to describe people's problem solving relative to this solution structure. These representations are similar to planning net representations discussed by VanLehn and Brown (1980).[3]

To illustrate, consider the planning nets for two of the Jasper episodes, *Journey to Cedar Creek* (JCC) and *Rescue at Boone's Meadow* (RBM; see Figures 1 and 2). In JCC, the main character, Jasper Woodbury, buys an old cabin cruiser. The cruiser's lights do not work, and it has a small, temporary gas tank. As depicted in Figure 1, the major plan that students need to evaluate is whether Jasper can drive his new cruiser home. Time and fuel are the constraints that must be tested to decide whether this plan is feasible. To determine whether Jasper has enough time and fuel to make it home, students must first determine the distance to Jasper's dock. Information about the distance to be traveled and the cruiser's average speed are used to determine an estimate of trip time. Once trip time is established, students can compare it with their estimate of the time available for the trip. Furthermore, the estimate of trip time can be combined with information about the cruiser's average fuel consumption rate to determine the amount of fuel that will be used during the trip. Students compare this with the amount of fuel that is available in the cruiser's tank to determine whether Jasper has a sufficient amount to make it home.

Jasper does have enough time, but does not have enough fuel to get home. The latter problem suggests a new plan, that of getting gasoline along the way. From the video, students know that gasoline is available at Willie's. Evaluating the feasibility of the plan involves determining whether Jasper has enough time and fuel to reach Willie's, and whether he has enough money to purchase the gasoline that he needs to make it home. Because the procedures for determining time and fuel to Willie are

---

[3]We thank James Greeno for bringing this to our attention.

Challenge: Can Jasper make it home before sunset?

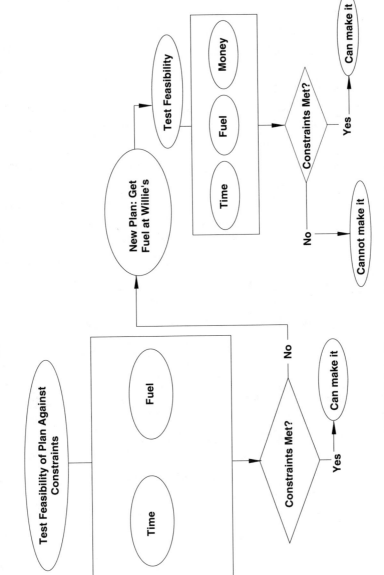

FIGURE 1. Planning net for *Journey to Cedar Creek*.

essentially the same as those used previously to determine time and fuel home, the planning net has iterative elements.

In RBM, Jasper goes on a fishing trip to Boone's Meadow and, while he is there, he finds a wounded eagle. He radios his friend Emily and asks for her help. As shown in Figure 2, Emily's challenge is to find the fastest way to rescue the eagle from the wilderness and to determine how long the rescue will take. There are multiple possible plans for rescuing the eagle that need to be tested against one another to find the fastest plan. As Figure 2 indicates, a plan involves determining a route (routes entail two, three, or four legs), vehicles to be used in the route, and agents who will be involved. Several rescue plans involve using a plane. If a plan involves a plane, then it must be tested against three constraints: payload, range, and landing space (these are irrelevant if the ultralight is not used in the rescue, i.e., if Jasper walks to Hilda and drives his car to Cumberland City). If payload or range is exceeded on any leg of a route that is tested, a new plan must be generated and its feasibility tested. This iterative process continues until at least two feasible plans have been generated. Execution times for each plan can then be compared (or the time for a single plan can be determined before generating and testing other plans).

The planning nets for JCC and RBM share some common features but contain some unique features as well. In both, students must test the feasibility of plans against a set of constraints, and some of the constraints are similar (e.g., fuel considerations). However, the JCC problem structure is less iterative than the structure for RBM from the perspective of the number of plans that need to be tested; fewer plans need to be generated and tested in JCC. The JCC solution is also more hierarchical in nature. There is a plan (getting gasoline at Willie's) embedded in another plan (can Jasper make it home), and, as such, there is a logical order in which to test plans. Finally, JCC and RBM differ in that RBM introduces optimization as a factor. In other words, students must go beyond finding a feasible rescue plan and find one that is optimum in the sense of requiring a minimum amount of time.

Baseline Studies of Problem Solving

As mentioned earlier, one of our goals has been to assess the degree to which typical school instruction prepares students to solve complex prob-

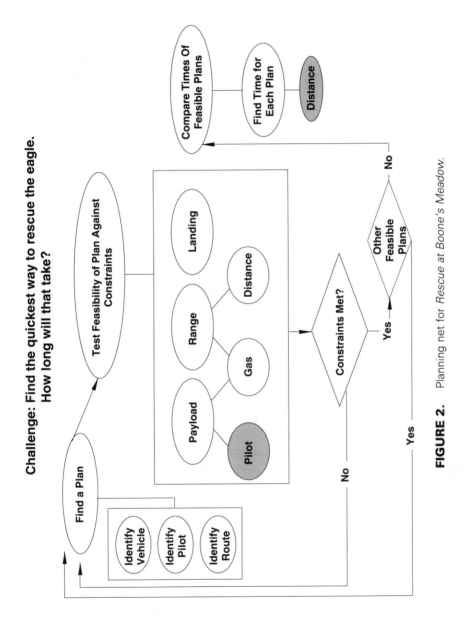

Challenge: Find the quickest way to rescue the eagle. How long will that take?

FIGURE 2. Planning net for *Rescue at Boone's Meadow*.

127

lems such as those found in Jasper. Therefore, we have conducted a series of *baseline* studies that assess people's abilities to solve Jasper problems prior to any instruction in these problems. Two types of students have participated in these baseline studies: (a) high-achieving sixth graders (who scored in stanines of 8 or 9 on mathematics achievement tests) and (b) college undergraduates. Our original reason for including undergraduates was to provide an expert group so that we could describe expert–novice differences in problem solving. However, as will be seen, even the undergraduates found the problems challenging, and their performance was far from perfect.

We used the idealized planning nets illustrated in Figures 1 and 2 as comparisons with actual planning nets generated by people during problem solving. Our goal was to describe problem solving in terms of (a) the accuracy with which a given subproblem is evaluated mathematically, (b) the scope of the solution space that is addressed by the problem solver (i.e., the number of plans and problem constraints that are considered), and, where relevant, (c) the degree to which problem solvers attempt to optimize their solutions. We report the results of baseline studies using the JCC and RBM videos. (For a more detailed reporting of the research findings, we refer readers to Goldman et al., 1992.)

The experimental protocol that we used for JCC and RBM was essentially the same. Students first viewed the Jasper episode, after which they were interviewed individually. The interviews were designed to assess student's abilities to formulate and to solve the challenge posed in the video and included a series of questions that provided students with progressively greater assistance in formulating the challenge. The initial question was general: The experimenter restated the challenge from the video. For JCC, students were asked, "Can Jasper get his boat home? Tell me as much as you can about all the problems he had to think about in order to make his decision, and solve these problems if you can." For RBM, the initial question was, "Emily has decided that she has come up with the quickest plan for rescuing the eagle. Tell me as much as you can about all the problems she had to solve to come up with the best plan, and solve these problems if you can." The remaining questions cued students to consider specific subproblems. For JCC, the second level of questions asked students to determine whether Jasper had enough time, fuel, and money to make the trip home. For RBM, the second level of questions asked students to consider the major constraints on route feasibility. Specifically, students were asked whether Emily needed to determine time for the rescue, fuel for the ultralight, and payload for the ultralight. JCC also included a third level, consisting of questions on how to solve each subproblem constituting the challenge. In all our studies, students were asked to talk aloud as they answered the questions. They were provided with paper, a pencil, and a summary of the Jasper story containing all of the data from the video. Interviews were audiotaped and subsequently transcribed for analysis.

Journey to Cedar Creek

Students' problem-solving protocols for JCC were scored in terms of the planning net elements illustrated in Figure 1, the goal being to describe their coverage of the solution space and their "mathematical success" in doing so. We examined the protocols for evidence that students had mentioned, attempted, or solved the major subgoals of the challenge (i.e., time to home, fuel to home, time to Willie's, fuel to Willie's, and money for gas). We computed a total score where 1 point was awarded for mentioning, attempting, and correctly solving each of these subgoals for a maximum of 15 points.

Analysis of the total scores indicated that both college students and sixth graders did relatively poorly, although college students had significantly higher scores than sixth graders. Figure 3 shows that scores increased at each level of questioning (i.e., Levels 1–3) for both groups. To explore this further, we looked at the percentage of students who mentioned, attempted, and solved each of the five subgoals. Figure 4 shows these percentages for Level 1 summarized across top-level versus embedded-level subgoals. The data indicate that roughly equal numbers of college students and sixth graders *mentioned* the top-level subgoals (i.e., mention either the time home or fuel home goals) but that more college students *attempted* and correctly *solved* these problems.

The data in Figure 4 also indicate that few students addressed the embedded goals portion of the planning net, that is, the time, fuel, and

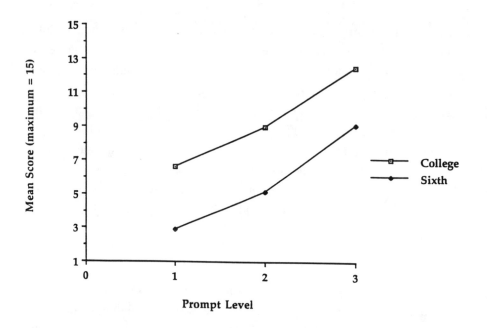

FIGURE 3.   Mean scores on solution space elements in JCC.

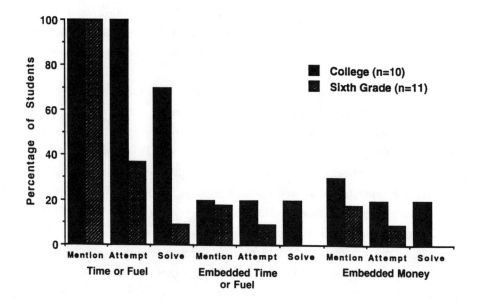

FIGURE 4.   Percentage mention, attempt, or solve for top-level and embedded goals in JCC (baseline).

money goals associated with the embedded plan of getting gasoline at Willie's. Far fewer students mentioned the embedded time or fuel goals than had mentioned these same goals in the top-level plan. (Consistent with these data is the fact that only a small handful of students in college or in the sixth grade mentioned money, a consideration that becomes important in determining whether there is a way to buy the needed fuel.) For the most part, if college students mentioned any of the embedded goals, they attempted and solved them successfully. However, none of the sixth graders solved any of the embedded goals.

Two important questions involve why students were less likely to explore the embedded goals and why the solution rates were so low. The data from the Level 2 and 3 prompting (see Figure 3) shed some light on these questions. At Level 2, college students were quite successful at solving the problems for which they were prompted (i.e., time home, fuel home, and money). However, Level 2 questions did not dramatically increase the number of college students who mentioned and solved the embedded goals of time to Willie's and fuel to Willie's. Their failure to explore these goals was probably not due to poor mathematical skills, because the embedded time and fuel problems entailed the same procedures as the top-level problems (i.e., time and fuel computations), and the Level 3 data indicated that most college students can successfully solve each of the goals when these are explicitly provided. Their failure to explore embedded goals may be due to a failure to generate or to think of the embedded plan in the first place (i.e., to see whether Jasper can get fuel at Willie's). Another possibility is that the college students may have been satisfied with producing a solution to a "more literal" interpretation of the challenge, that is, to determine whether Jasper had sufficient time and fuel to make it home without trying to pursue alternate plans, such as how he could get more fuel.

Why did few of the sixth graders explore embedded goals? The reasons appear to be mixed. Part of their difficulty appears to be mathematical because, even at Level 3, not all students solved the subproblems. They also appear to have planning or organizational difficulties because they perform better on top-level goals at Level 3 than they do at Level 2 or 1. In general, the sixth graders do much better when subgoals are made explicit for them (Level 3) then when they are left implicit (Levels 1 and 2). This suggests that, even though students may have the mathematical knowledge necessary to solve particular problems, they also need skills for representing complex problems, especially skills that require them to *generate* and to coordinate the solving of their own subgoals.

Rescue at Boone's Meadow

As with JCC, in analyzing the content of the RBM protocols, we looked at whether students mentioned, attempted to solve, or successfully solved a plan by testing the constraints imposed by range, payload, and time factors. In addition, we also examined whether students considered more than one plan (i.e., route or vehicle) or compared the times for two rescue plans, considerations necessary for optimizing the solution. For example, imagine that a student devised one plan for a rescue (e.g., have Jasper carry the wounded eagle to Hilda's and go from there by car) and figured out the time that it would take to execute this plan. We looked at the degree to which students stopped after formulating a plan, or at whether they evaluated the time that it would take to execute it and thought about other plans that might be more optimal (e.g., using the ultralight to land at Boone's Meadow and pick up the eagle).

The mention/attempt/solve data for RBM were similar to these same data for the top-level goals in JCC. (In view of the fact that RBM does not contain an embedded plan, it was expected that if there were any similarities with JCC, they could occur only with JCC top-level goals.) In other words, most sixth-grade and college students mentioned the major goals (i.e., range, time, or payload), but fewer sixth graders than college students attempted and solved these goals. The exception to this pattern was the payload data: Equal numbers of college and sixth-grade students attempted and solved this goal. The lack of a group difference on payload is probably attributable to the fact that the mathematics needed to solve the problem is simple addition and thus is relatively less difficult for sixth graders than the mathematics needed to solve the range and time problems.

The data on optimization were particularly interesting. We compared the number of students who generated one plan versus more than one plan for rescuing the eagle. We also compared the number of students who determined and compared times associated with their alternative plans. As reflected in the attempt data, college students (89%) were more likely than sixth graders (40%) to try to figure out how long it would take to rescue the eagle using the plan or plans that they had generated. However, group differences were much smaller in the extent to which they developed optimal plans: Fifty percent of the sixth-grade students and 45% of the college students considered only *one* plan for rescuing the eagle. Furthermore, of the sixth-grade students who generated multiple plans, 50% attempted to determine time, but only half of these students compared times for alternative plans. Among the college students who generated multiple plans, 80% determined time, but only half of these students compared times. The low rate of optimization is particularly interesting because the challenge explicitly asked them to find the *quickest* way to rescue the eagle. These results suggest that neither group was very successful at optimizing the solution.

Discussion of the Baseline Data

In summary, the baseline data for JCC and RBM suggest that both sixth-grade and college students were relatively good at identifying the constraints that must be tested to evaluate their solution plans, although sixth graders were less likely to try to solve and to solve successfully these problems. The results from RBM also indicate that students, especially sixth graders, tend not to try to optimize their solutions by considering and evaluating multiple plans. Finally, the results from JCC suggest that problems that entail embedded plans are not solved very well by either sixth-grade or college students, although the reasons why the groups perform poorly may differ. Sixth graders appear to lack some important mathematical and general problem-solving skills.

Effects of Instruction on Transfer

The findings from the baseline studies suggest that the Jasper problems are challenging with regard to the planning and mathematical skills that are needed for solving them. An important issue raised by the baseline research is the extent to which students' problem-solving performance can be improved with instruction on Jasper, instruction that emphasizes the generation and definition of plans and goals, as well as mathematical skill. As discussed earlier in this chapter (see also CTGV, 1992), Jasper instruction, as we envision it, is generative in nature. Under the guidance of the instructor, students generate the solution plans and subgoals, search for relevant data, and work cooperatively in small groups to find solutions. We have conducted several studies contrasting the effects of this form of instruction with a method derived from the work of Polya (1957) in which traditional word problems of the type needed for the JCC solution served as the instructional context.

Transfer Study 1

In our first study, participants were fifth graders from a high-achieving mathematics class. Students were assigned within class to Jasper instruction or to word-problem instruction groups. Before and after their respective instruction (instruction took place in four 1-hour sessions), groups were given tests to assess learning and transfer. One test was designed to tap students' mastery of the solution to JCC and consisted of items on which students had to match subgoals with their more superordinate goals and subgoals with the data that were relevant for solving them. Not surprisingly, students in the Jasper group scored much better on the mastery test. For present purposes, we focus on the results of a near-transfer test administered to students after the instruction because they relate directly to the results of the baseline studies discussed earlier. For a discussion of the complete results, we refer the reader to Goldman, Vye, Williams, Rewey, and Pellegrino (1991); Goldman and the CTGV (1991); and Van Haneghan et al. (1992).

The transfer test was designed to assess transfer from JCC to a highly similar problem. The transfer problem was video based and its solution was isomorphic to the JCC challenge. It told the story of a character named Nancy who buys a houseboat and must then decide whether she can get the boat home before sunset without running out of fuel. The boat has the same problems as Jasper's cruiser in JCC (it has a small temporary fuel tank, and its running lights do not work). The solution has the same structure as JCC, although the data are different; it involves consideration of the same subproblems and has the same outcomes. Students watched the transfer video and then solved the problem while talking aloud. The interview procedures were the same as those used in the baseline studies. The initial question was general and asked students to decide whether Nancy could make it home (i.e., Level 1); subsequent Level 2 questions prompted them to consider three of the subgoals (time home, fuel home, and money).

As in the JCC baseline study, students' problem-solving protocols were scored in terms of the types of planning net elements contained in Figure 1. Once again, we examined the protocols for evidence that students had mentioned, attempted, or solved the major subproblems of the challenge (i.e., time to home, fuel to home, time to Tom's, fuel to Tom's, and money for gasoline). The analyses of the total scores indicated that Jasper students achieved higher scores than the word-problem students. (Level 2 questions were associated with the higher scores than the Level 1 question, but this outcome did not interact with the group.) To explore the data further, we looked at the percentage of students who mentioned, attempted, and solved each of the five subproblems. Figure 5 shows these percentages for Level 1 summarized for top-level and embedded goals.

Results indicate large differences between the Jasper group and the word-problem group. More than 75% of the Jasper-instructed group solved at least one of the top-level goals, compared with less than 20% of the word-problem groups. As was true in our baseline study, the embedded time and fuel goals were considered and solved by fewer students than the top-level time and fuel goals, but the embedded goal percentages were greater for Jasper students than for word-problem students. (Once again the embedded money goal is reported separately to allow direct comparison of the time and fuel goals at the top and embedded plan levels). The major effect of the Level 2 prompting was to increase the percentage of Jasper students solving the top-level goals; the numbers of students attempting and solving the embedded goals remained relatively constant from Levels 1 to 2. Overall, the performance after instruction of the fifth graders in the Jasper group was as good as the college students in the JCC baseline study who had not received instruction in Jasper. Further-

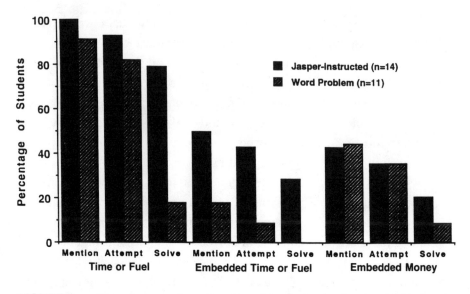

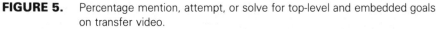

FIGURE 5. Percentage mention, attempt, or solve for top-level and embedded goals on transfer video.

more, the word-problem group showed an uncanny resemblance to the performance of the sixth-grade baseline students.

It is also important to note that we gave all our students posttests on one- and two-step word problems similar to those practiced by the word-problem group, word problems that involved the same basic distance–rate–time concepts as the Jasper problem. Both groups did very well on these problems. However, a number of the students in both groups were at ceiling on the test, making it difficult to draw firm conclusions about how much the groups had learned.

Transfer Study 2

We used the preceding instructional design (i.e., Jasper instruction versus word-problem instruction) in a second study that involved a different test of transfer. The transfer problem was the Jasper video RBM. In contrast to the Nancy transfer problem, RBM does not share specific mathematical procedures or the same goal organization as JCC, but RBM and JCC do share elements of a general trip planning schema. For example, several of the constraint-testing elements from JCC, specifically time and fuel considerations, are relevant to the RBM. We predicted positive transfer on those elements of the schema that overlap from the first to the second trip-planning adventure. On the other hand, the optimization elements of RBM are not present in the JCC adventure. For this aspect of problem solving, we did not expect to see positive transfer from JCC to RBM.

As in Transfer Study 1, high-achieving fifth graders were randomly assigned to either a Jasper or a word-problem instruction group. The lesson plans used in Transfer Study 1 were used for each group. Again, prior to and following instruction, students were administered a series of tests designed to test mastery of JCC and transfer. For present purposes, we focus on the transfer results, although we note that Jasper-instructed students performed significantly better on these instruments than word-problem-instructed students.

The RBM transfer test was administered using the same protocol as the RBM baseline test. We analyzed these interviews by looking for elements associated with finding and testing feasible routes (i.e., range, payload, landing, and time) as well as elements associated with optimizaton of routes (i.e., were multiple routes or vehicles proposed and were time estimates compared). When we computed a total score for each student, we found that Jasper students scored significantly higher than word-problem students, indicating that Jasper students explored more of the RBM solution space than word-problem students. Again, we examined this effect further by looking at the percentages of students who mentioned, attempted, and solved the feasibility constraints. These data are presented in Figure 6. We expected that Jasper students would show positive transfer from JCC to RBM on the time and range subproblems. The range subproblem shows evidence of this effect in terms of the number of students attempting and solving. The time problem does not, although the only student to solve the time problem successfully was in the Jasper-instructed condition. In addition to these specific transfer effects, data also indicated that a greater number of Jasper-instructed students mentioned, attempted, and solved the payload problem. Because issues of payload were unique to RBM, these data suggest that Jasper-instructed students may have learned a general heuristic from JCC (i.e., generate possible constraints on plans and test against them).

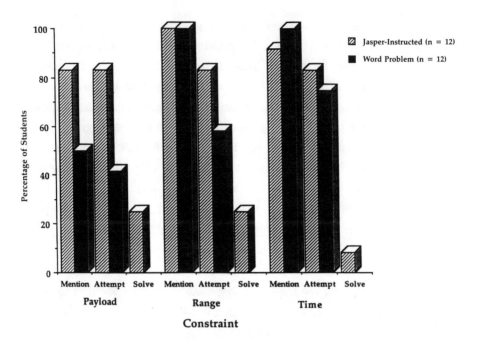

FIGURE 6. Jasper versus word-problem instruction: transfer to *Rescue at Boone's Meadow* constraint inclusion.

We also looked at the extent to which students attempted to optimize their solution. As noted earlier, we did not expect to see group differences in this aspect of problem solving because JCC instruction does not focus on generating and evaluating multiple plans for purposes of optimization. The results confirmed this expectation: Thirty-three percent of the Jasper students and 41% of the word-problem students generated only one plan for rescuing the eagle. Furthermore, although most students in both groups tried to determine how much time their plan or plans would take, in both groups, less than half of the students who generated more than one plan compared the time estimates associated with their plans to decide which was fastest.

Discussion of the Transfer Data

A potential problem with the transfer tasks used in the two preceding studies is that they were visual in nature. Perhaps the positive effects on transfer to complex problems is due simply to the fact that Jasper students become familiar with video-based problems and control students do not. There are at least three reasons to question the validity of this argument. First, as noted in the last study, evidence of positive transfer tended to be systematically related to those parts of the transfer problem that were similar to the kinds of learning experiences available previously. These data suggest that something other than mere "familiarity with video-based problems" is going on. Second, in both of the previously discussed transfer studies, the word-problem group saw a Jasper at the beginning of their instruction—they simply did not solve it. So, they were familiar with the format that told a story and ended with a problem to solve. In addition, for the transfer test, students first saw a video, but were then given written summaries of what they had seen. They were able to work from these written summaries, so this format was familiar to the word-problem group. We did not rely exclusively on written transfer problems because our experience indicates that sole use of the written format is a cumbersome way to present problems that are as complex as the ones that we used in our transfer tests.

A third reason for questioning the argument that our evidence for transfer is due only to the fact that Jasper students are familiar with video problems is that we have conducted other studies that make exclusive use of written problems that are more complex than one- and two-step problems (e.g., see CTGV, 1992d; Goldman et al., 1992). These written problems have still been less complex than the ones used in the studies discussed previously because, as noted earlier, it becomes cumbersome to attempt to communicate such complexity by making sole use of the verbal medium. Nevertheless, the problems used are complex enough to convince us that transfer occurs for reasons other than mere familiarity with visually based problems.

Overall, there are several findings from the studies described earlier that seem noteworthy. First, our baseline studies indicate that, without instruction, sixth-grade students who score very high on traditional tests of mathematics achievement have a very difficult time solving complex problems such as those found in Jasper. Even college students have more difficulty with Jasper problems than we expected initially. These baseline studies also show that there are important differences between the ability to solve well-defined one- and two-step problems that underlie the solution to Jasper problems (the problems provided at Level 3 in our baseline interviews on JCC) and the ability to begin with a complex scenario (such as a Jasper) and generate the subgoals and subproblems that one needs to solve to accomplish an overall goal.

A second major finding of our work stems from the transfer studies indicating that instruction that focuses on general problem-solving strategies (i.e., Polya's heuristics) in the context of solving the one- and two-step equivalents of Jasper problems is not sufficient to help students to learn to deal with complex Jasper-like problems that require them to generate their own subgoals and proceed from there.

A third major finding is that transfer from solving a Jasper adventure (such as JCC) seems to involve some general and some specific factors. Transfer data from JCC to RBM indicate that the strongest transfer occurs for those elements of the RBM problem (e.g., determining fuel constraints) that are similar to those found in the JCC problem. However, some transfer also occurs for new elements of the problem, such as the need to consider payload constraints of the ultralight used in RBM. In general, we expect transfer to improve even further to the extent that students receive instruction in a *series* of Jaspers and are helped to represent these experiences in ways that are applicable to a wide variety of settings. This hypothesis about the effects of a series of adventures and about the value of helping students to create meaningful and general representations of these experiences awaits further research.

Taken as a whole, our findings suggest optimism about the value of anchored problem-solving experiences such as those found in the Jasper Series. Nevertheless, we have begun to ask ourselves whether more is involved in transfer than the mere ability to solve new problems that are similar to those solved previously. As indicated in the next section, our attempts to probe further the students' understanding have led us to modify our thinking about assessing learning and to modify our Jasper materials as well.

Some New Concerns About Learning and Transfer

As we began to discuss Jasper adventures among ourselves, it gradually became clear that there is a type of problem-solving expertise that is different from what is measured by asking whether people can transfer to problems that are different from, yet analogous to, problems solved earlier. We found that as we became familiar with Jasper's world and attempted to solve problems in it, it became natural to use the results of previous problem-solving experiences to provide shortcuts to the solution of new problems. For example, imagine being asked to predict the time that it would take Jasper to get from Cedar Creek to his home dock in a new boat that had a 24-gallon fuel tank, burned 5 gallons per hour, and cruised at a speed of 16 mph. A novice in Jasper's world would have to ask for information about the distance from Cedar Creek to Jasper's home dock and to do a number of calculations to determine the time for the journey and whether Jasper would have enough gas to make it. In contrast, an expert in Jasper's world could take many shortcuts. If she or he remembered that Jasper's original boat traveled at a rate of 8 mph and took 3 hours for the journey, it is an easy matter to determine that the new boat would take only 1 1/2 hours for the journey. In addition, because Jasper's original boat had a 12-gallon tank, burned at 5 gallons per hour at cruising speed, and was able to make the trip by stopping once for gasoline, it is clear that the new boat could also make the trip.

We refer to problems such as the one just described as "what-if" analog problems. They perturb variables in previously solved problems rather than present students with problems that are totally new yet analogous in structure to ones solved before. Recently, we have begun to explore students' performance on what-if problems. Thus, we might ask them to consider the effects on Jasper's trip if the cruiser had a 13-gallon fuel tank or if its speed were 16 mph instead of 8 mph. We have compared the performance of two groups of students on these what-if problems: Students who have solved a Jasper adventure and those who have seen the Jasper adventure but not solved it.

At one level, students who have solved an original Jasper problem should have an advantage over students in control groups. The Jasper students should be able to use declarative knowledge about the results of their previous computations to take shortcuts. But the ability to take shortcuts also requires knowledge about which sets of declarative knowledge from previous attempts at problem solving apply to the new what-if problems. The ability to decide when to use previously acquired declarative knowledge and when to recompute answers may be problematic for students, especially because most curricula are not anchored and hence do not present students with new problems that are perturbations on earlier problems that they have already solved.

In our study (Williams et al., 1992), Jasper students and control students saw JCC first. The Jasper students received instruction in JCC, whereas the control students worked in their traditional mathematics curriculum. After JCC instruction, we gave all students a written test designed to assess their mastery of the JCC solution and then gave them individual interviews on what-if analog problems. The results of the JCC mastery test indicated that Jasper-instructed students outperformed the control students and that this was the case for both the high-achieving and average-achieving students who participated. Based on their performance on the JCC mastery tests, experimental students were expected to have acquired enough knowledge about JCC to be able to use it when solving the what-if analogs.

The what-if problem discussed in this chapter was generated by increasing the speed of the cruiser in JCC from 8 to 9 mph. Students were asked to decide whether Jasper could make it home before dark without running out of fuel if he traveled at a rate of 9 mph. Changing the speed of the cruiser affects trip time (Jasper gets home faster) and fuel consumption because, in JCC, fuel consumption is given in gallons burned per hour at cruising speed. Assuming that fuel consumption remains invariant, Jasper's ability to get home faster means that he will not use as much gasoline as he did in the original adventure.

Students were asked to talk aloud while solving the analog. All students were supplied with an illustrated storyboard giving the original facts from JCC, except that the speed of the boat was changed to 9 mph.

The protocols from students solving the analog problem were analyzed in terms of the number of subgoals (i.e., all 16 subproblems needed to solve the challenge) attempted and the type of reasoning used by students. Results indicated that Jasper students addressed more of the problem space when solving the analog; they attempted to solve significantly more subgoals than control students. For example, Jasper students were more likely than control students to consider whether they had enough money to buy gasoline at Willie's.

We also analyzed the protocols for the types of reasoning used to solve each subgoal. For present purposes, we focus on two general categories of reasoning: (a) reasoning from prior context that makes use of previous knowledge from JCC and (b) reasoning from present data that involves calculation of answers without reference to JCC. Given the what-if problem of a 9 mph cruising speed, protocols were scored as making reference to JCC if they included evidence of recalling various outcomes from JCC (e.g., it took Jasper 3 hours to get home at 8 mph) and using this information when thinking through the analog problem (e.g., therefore, at 9 mph, Jasper should make it home faster than at 8 mph). A solution method that does not involve reference to JCC is to use the storyboard data sheets handed out as part of the what-if problems to calculate the answer. Thus, at 9 mph, the time for Jasper to make it home would be 24 miles (the distance from Cedar Creek to Jasper's home dock) divided by 9 mph.

Table 1 shows the percentage of responses given by students in the Jasper and control groups to what-if questions that spontaneously used information from their previous work with JCC. (Although students in the control group had seen JCC but had not solved it, it is still possible for them to reason on the basis of information from JCC.) The data indicate that Jasper-instructed students frequently used outcome information from JCC. It is important to note, however, that Table 1 codes type of reasoning irrespective of whether it was right or wrong. Further analyses of the data indicate that, at least half of the time, Jasper students retrieved outcomes from JCC and applied them inappropriately to the

TABLE 1

Percentage of Responses That Used Information From JCC

Group	High achieving	Average achieving
Experimental (JCC)	37%	44%
Control[a]	6%	2%

[a]The nonzero data reflect the fact that control students occasionally used outcome information from the mastery test they had completed prior to the analogs.

solution of the analog. For example, students might state that it will cost Jasper the same amount to buy gas as it did in the original JCC. Because fuel consumption is set at 5 gallons per hour at cruising speed and the cruiser goes faster in the what-if analog problem, Jasper would actually need less gasoline in the analog and, hence, it would cost him less.

Although there are many additional dimensions to the what-if analog study, the data discussed earlier are sufficient to make several points that we want to emphasize. First, it was interesting to us that fifth-grade students who had experience in solving JCC spontaneously attempted to make use of declarative knowledge from previous computations (e.g., that it took Jasper 3 hours to get home) in order to deal with what-if perturbations of the original problems. The students did this despite the fact that the Jasper instruction had never exposed them to what-if analog problems. Prior to running the study, we had noticed our own tendencies to spontaneously use the outcomes of previous computations to reason about possible what-if extensions to various Jasper adventures, but we had no idea whether this type of reasoning would be natural for fifth and sixth graders. Our data suggest that it is.

On the other side of the coin, our data suggest that students often made mistakes when attempting to use previously acquired knowledge to reason about the what-if problems. Knowing when to use the results of previous outcomes and when to recompute them can be a complex task. It requires a deep understanding of relationships among various parts of the problem. For example, the ability to think about the 9-mph cruising speed, what-if analog by comparing it to the original JCC adventure (which involved an 8-mph cruising speed) requires an understanding of how speed affects fuel consumption.

Overall, our work with what-if analogs has helped us to identify a problem with attempts to measure learning by relying exclusively on transfer problems that involve totally new content yet are structurally analogous to ones solved earlier (e.g., see our earlier discussion of our Nancy and RBM transfer tests as well as Gick & Holyoak, 1980, 1983). It now seems clear to us that students could learn the procedures for solving a Jasper adventure such as JCC, and could then reapply these procedures to a transfer test such as our Nancy houseboat transfer or RBM, yet could still fail to have developed a deep understanding of relationships among variables in the problem. The use of what-if analog problems that perturb variables in previously solved problems appears to be a way to assess more deeply the degree to which students understand.

Changes in the Jasper Materials

Our what-if analog studies have prompted us to change the design of our Jasper adventures. After solving the major problem for each adventure, we now encourage teachers to have groups of students work on what-if analog and extension problems. The purpose of these materials is to help students to develop flexible knowledge representations, to better understand key mathematical principles embedded in the Jasper adventures, and to make connections between the adventures and the thinking and planning that took place in many historical and contemporary events. Relevant video-based analogs and extensions to each problem now appear on each Jasper videodisc.

Analog problems are formed by altering one or more of the parameters of the original Jasper problem. For example, after students have solved JCC, which took place on a day on which the river current was only 1 mph, they can be asked to reconsider Jasper's trip time if the dam had been opened and the current was 3 mph (Jasper was traveling downstream). Similarly, students can be asked to imagine that the cruiser's speed was 9 mph, and so on. We are also designing an analog "adventure maker" that allows students to create their own analog adventures given specific constraints. For example, students might be asked to choose among a set of values that would allow Jasper to get home before sunset if his boat had only an 8-gallon rather than a 12-gallon temporary tank.

In addition to adding what-if analog problems to each Jasper disc, we decided to include others that would allow us to accomplish additional sets of goals. Thus, some of our analog problems are specifically designed to help students explore important concepts in mathematics. As an illustration, students may be asked to imagine that the cruiser in JCC had a gas tank with length, width, and height dimensions such as $2n \times 2n \times 2n$ rather than $n \times n \times n$. Would the new tank hold twice as much gasoline? Problems such as these can help students to explore concepts such as scaling factors and their effects on perimeter versus area versus volume.

Extension problems are designed to help students to integrate their knowledge across the curriculum, for example, to see how the planning involved in the Jasper adventures relates to historical and current events. One extension problem for JCC involves a tugboat pushing barges down the Mississippi River from St. Louis to New Orleans. Students are asked to generate what the captain needs to think about to plan the trip. They are then given relevant data and asked to determine the estimated time for the trip and amount of fuel needs.

Summary and Conclusion

Our goals in this chapter were fourfold: (a) provide a theoretical overview of our center's approach to anchored instruction; (b) describe an example of our approach, the Jasper Woodbury problem-solving series; (c) discuss our overall research program and one subset of this program that focuses on studies of problem representation and transfer; and (d) discuss some of the changes in materials design that have been motivated by our research.

Programs such as Jasper both draw on and contribute to the psychological literature on thinking, learning, and problem solving. We drew on the psychological literature on learning to design the initial adventures, which, in turn, has allowed us to study complex problem solving by people ranging from fifth graders to adults. In the process, we have deepened our understanding of issues of learning and transfer. For example, the use of Jasper problems allows us to study problem solving that is both complex and realistic; the use of the visual medium makes it possible to do with relatively young students because the video helps make the complexity manageable. In addition, we noted that most studies in the psychological literature assess transfer by asking people to solve problems that include new content but analogous solution structures. We, too, use these types of transfer problems, but have begun to see the advantages of additional types as well. In particular, the use of what-if problem-solving analogs allows us to assess the degree to which people spontaneously attempt to use the products of previous computations as shortcuts for solving subsequent problems that perturb variables in the original problem. Problems such as these seem to be encountered frequently in real life (e.g., we might try to estimate expenses for a trip that we took before, except that we now have a new car that gets twice the gasoline mileage). Our future research will include assessments of the effects of working with what-if analogs on students' abilities to understand more deeply problems that they have been asked to solve.

References

Adams, L., Kasserman, J., Yearwood, A., Perfetto, G., Bransford, J., & Franks, J. (1988). The effects of facts versus problem-oriented acquisition. *Memory & Cognition, 16*, 167–175.

Barron, B. J. S. (1991). *Collaborative problem solving: Is team performance greater than what is expected from the most competent member?* Unpublished doctoral dissertation, Vanderbilt University, Nashville, TN.

Barrows, H. S. (1985). *How to design a problem-based curriculum for the preclinical years.* New York: Springer.

Bransford, J. D., Franks, J. J., Morris, C. D., & Stein, B. S. (1979). Some general constraints on learning and research. In L. S. Cermak & F. I. M. Craik (Eds.), *Levels of processing and human memory* (pp. 331–354). Hillsdale, NJ: Erlbaum.

Bransford, J. D., Franks, J. J., Vye, N. J., & Sherwood, R. D. (1989). New approaches to instruction: Because wisdom can't be told. In S. Vosniadou & A. Ortony (Eds.), *Similarity and analogical reasoning* (pp. 470–497). Cambridge, England: Cambridge University Press.

Bransford, J. D., Goldman, S. R., & Vye, N. J. (1991). Making a difference in peoples' abilities to think: Reflections on a decade of work and some hopes for the future. In L. Okagaki & R. J. Sternberg (Eds.), *Directors of development: Influences on children* (pp. 147–180). Hillsdale, NJ: Erlbaum.

Bransford, J. D., Sherwood, R., & Hasselbring, T. (1988). The video revolution and its effects on development: Some initial thoughts. In G. Foreman & P. Pufall (Eds.), *Constructivism in the computer age* (pp. 173–201). Hillsdale, NJ: Erlbaum.

Bransford, J. D., Sherwood, R. S., Hasselbring, T. S., Kinzer, C. K., & Williams, S. M. (1990). Anchored instruction: Why we need it and how technology can help. In D. Nix & R. Spiro (Eds.), *Cognition, education, and multi-media: Exploring ideas in high technology* (pp. 115–141). Hillsdale, NJ: Erlbaum.

Bransford, J. D., & Vye, N. J. (1989). A perspective on cognitive research and its implications for instruction. In L. Resnick & L. E. Klopfer (Eds.), *Toward the thinking curriculum:*

Current cognitive research (pp. 173–205). Alexandria, VA: Association for Supervision and Curriculum Development.

Brown, J. S., Collins, A., & Duguid, P. (1989). Situated cognition and the culture of learning. *Educational Researcher, 18*(1), 32–41.

Charles R., & Silver, E. A. (Eds.). (1988). *The teaching and assessing of mathematical problem solving.* Hillsdale, NJ: Erlbaum & National Council of Teachers of Mathematics.

Chi, M. T., Bassok, M., Lewis, P. J., & Glaser, R. (1989). Self-explanations: How students study and use examples in learning to solve problems. *Cognitive Science, 13*, 145–182.

Clement, J. (1982). Algebra word problem solutions: Thought processes underlying a common misconception. *Journal of Research in Mathematics Education, 13*, 16–30.

Cobb, P., Yackel, E., & Wood, T. (1992). A constructivist alternative to the representational view of mind in mathematics education. *Journal for Research in Mathematics Education, 23*, 2–33.

Cognition and Technology Group at Vanderbilt. (1990). Anchored instruction and its relationship to situated cognition. *Educational Researcher, 9*(6), 2–10.

Cognition and Technology Group at Vanderbilt. (1991). Technology and the design of generative learning environments. *Educational Technology, 31*(5), 34–40.

Cognition and Technology Group at Vanderbilt. (1992a). Anchored instruction and science and mathematics: Theoretical basis, developmental projects, initial research findings. In R. Duschl & R. Hamilton (Eds.), *Philosophy of science, cognitive psychology, and educational theory and practice* (pp. 245–273). New York: SUNY Press.

Cognition and Technology Group at Vanderbilt. (1992b). The Jasper experiment: An exploration of issues in learning and instructional design. *Educational Technology Research and Development, 40*(1), 65–80.

Cognition and Technology Group at Vanderbilt. (1992c). The Jasper Series: A generative approach to improving mathematical thinking. In S. M. Malcom, L. G. Roberts, & K. Sheingold (Eds.), *This year in school science* (pp. 109–140). Washington, DC: American Association for the Advancement of Science.

Cognition and Technology Group at Vanderbilt. (1992d). The Jasper Series as an example of anchored instruction: Theory, program description and assessment data. *Educational Psychologist, 27*, 291–315.

Cognition and Technology Group at Vanderbilt. (1993). Toward integrated curricula: Possibilities from anchored instruction. In M. Rabinowitz (Ed.), *Applied cognition: Developing instruction and assessing competence* (pp. 33–55). Hillsdale, NJ: Erlbaum.

Cognition and Technology Group at Vanderbilt. (in press). Integrated media: Toward a theoretical framework for utilizing their potential. *Journal of Special Education Technology.*

Cosden, M. A., Goldman, S. R., & Hine, M. S. (1990). Learning handicapped students' interactions during a microcomputer-based writing activity. *Journal of Special Education Technology, 10,* 220–232.

Dewey, S. (1933). *How we think: Restatement of the relation of reflective thinking to the educative process.* Boston: Heath.

Duffy, T. M., & Bednar, A. K. (1991). Attempting to come to grips with alternative perspectives. *Educational Technology, 31*(9), 12–15.

Gibson, J. J. (1977). The theory of affordance. In R. Shaw & J. Bransford (Eds.), *Perceiving, acting, and knowing* (pp. 67–82). Hillsdale, NJ: Erlbaum.

Gick, M. L., & Holyoak, K. J. (1980). Analogical problem solving. *Cognitive Psychology, 12,* 306–365.

Gick, M. L., & Holyoak, K. J. (1983). Schema induction and analogical transfer. *Cognitive Psychology, 15,* 1–38.

Goldman, S. R., Cosden, M. A., & Hine, M. S. (1987, April). *Writing groups: A context for interactive communication among learning disabled children.* Poster presented at the meeting of the Society for Research in Child Development, Baltimore, MD.

Goldman, S. R., Cosden, M. A., & Hine, M. S. (1992). Working alone and working together: The relationship between grouping and characteristics of learning handicapped students' stories. *Learning and Individual Differences, 4,* 369–393.

Goldman, S. R., Pellegrino, J. W., & Bransford, J. D. (in press). Assessing programs that invite thinking. In E. Baker & H. Oill (Eds.), *Technology assessment: Estimating the future.* Hillsdale, NJ: Erlbaum.

Goldman, S. R., Vye, N. J., Williams, S. M., Rewey, K., & Pellegrino, J. W. (1991, April). *Problem space analyses of the Jasper problems and students' attempts to solve them.* Paper presented at the American Educational Research Association, Chicago, IL.

Goldman, S. R., & the Cognition and Technology Group at Vanderbilt. (1991, August). *Meaningful learning environments for mathematical problem solving: The Jasper problem solving series.* Paper presented at the Fourth European Conference for Research on Learning and Instruction, Turku, Finland.

Gragg, C. I. (1940). Because wisdom can't be told. *Harvard Alumni Bulletin,* 78–84.

Hanson, N. R. (1970). A picture theory of theory meaning. In R. G. Colodny (Ed.), *The nature and function of scientific theories* (pp. 233–274). Pittsburgh, PA: University of Pittsburgh Press.

Hickey, D. T., Pellegrino, J. W., Petrosino, A., & the Cognition and Technology Group at Vanderbilt. (1991, October). *Reconceptualizing space science education: A generative, problem solving approach.* Paper presented at the Florida Space Education Conference, Cocoa Beach.

Hine, M. S., Goldman, S. R., & Cosden, M. A. (1990). Error monitoring by learning handicapped students. *Journal of Special Education, 23,* 407–422.

Jenkins, J. J. (1979). Four points to remember: A tetrahedral model and memory experiments. In L. S. Cermak & F. I. M. Craik (Eds.), *Levels of processing in human memory* (pp. 429–446). Hillsdale, NJ: Erlbaum.

Lamon, M. (1992, April). *Learning environments and macrocontexts: Using multimedia for understanding mathematics.* Paper presented at the annual meeting of the American Educational Research Association, San Francisco.

Lockhart, R. S., Lamon, M., & Gick, M. L. (1988). Conceptual transfer in simple insight problems. *Memory & Cognition, 16,* 36–44.

McLarty, K., Goodman, J., Risko, V., Kinzer, C. K., Vye, N., Rowe, D., & Carson, J. (1990). Implementing anchored instruction: Guiding principles for curriculum development. In J. Zutell & S. McCormick (Eds.), *Literacy theory and research: Analyses from multiple paradigms* (pp. 109–120). Chicago, IL: National Reading Conference.

Minstrell, J. A. (1989). Teaching science for understanding. In L. B. Resnick & L. E. Klopfer (Eds.), *Toward the thinking curriculum: Current cognitive research* (pp. 129–149). Alexandria, VA: Association for Supervision and Curriculum Development.

National Council of Teachers of Mathematics. (1989). *Curriculum and evaluation standards for school mathematics.* Reston, VA: Author.

Nickerson, R. S. (1988). On improving thinking through instruction. *Review of Research in Education, 15,* 3–57.

Palincsar, A. S., & Brown, A. L. (1984). Reciprocal teaching of comprehension fostering and comprehension monitoring activities. *Cognition and Instruction, 1,* 117–175.

Palincsar, A. S., & Brown, A. L. (1989). Instruction for self-regulated reading. In L. B. Resnick & L. E. Klopfer (Eds.), *Toward the thinking curriculum: Current cognitive research* (pp. 19–39). Alexandria, VA: Association for Supervision and Curriculum Development.

Pellegrino, J. W., Hickey, D., Heath, A., Rewey, K., Vye, N. J., & Cognition and Technology Group at Vanderbilt. (1991). *Assessing the outcomes of an innovative instructional program: The 1990–1991 implementation of the "Adventures of Jasper Woodbury"* (Tech. Rep. No. 91-1). Nashville, TN: Vanderbilt University, Learning Technology Center.

Perkins, D. N. (1991). What constructivism demands of the learner. *Educational Technology, 31*(9), 19–21.

Polya, G. (1957). *How to solve it.* Princeton, NJ: Princeton University Press.

Porter, A. (1989). A curriculum out of balance: The case of elementary school mathematics. *Educational Researcher, 18*(5), 9–15.

Resnick, L. (1987). *Education and learning to think.* Washington, DC: National Academy Press.

Resnick, L. B., & Klopfer, L. E. (Eds.). (1989). *Toward the thinking curriculum: Current cognitive research.* Alexandria, VA: Association for Supervision and Curriculum Development.

Resnick, L. B., & Resnick, D. P. (1991). Assessing the thinking curriculum: New tools for educational reform. In B. Gifford & C. O'Connor (Eds.), *Changing assessments: Alternative views of aptitude, achievement, and instruction* (pp. 37–76). Norwell, MA: Kluwer Academic.

Rewey, K. L., & the Cognition and Technology Group at Vanderbilt University. (1991, October). *Scripted cooperation and anchored instruction: Interactive tools for improving mathematics problem solving.* Paper presented at the McDonnell Meetings of the Cognition in Education Program, Nashville, TN.

Salomon, G., & Globerson, T. (1989). When teams do not function the way they ought to. *International Journal of Educational Research, 13,* 89–99.

Scardamalia, M., & Bereiter, C. (1985). Fostering the development of self-regulation in children's knowledge processing. In S. F. Chipman, J. W. Segal, & R. Glaser (Eds.), *Thinking and learning skills: Research and open questions* (Vol. 2, pp. 563–578). Hillsdale, NJ: Erlbaum.

Scardamalia, M., & Bereiter, C. (1991). Higher levels of agency for children in knowledge building: A challenge for the design of new knowledge media. *Journal of the Learning Sciences, 1,* 37–68.

Schoenfeld, A. H. (1985). *Mathematical problem solving.* San Diego, CA: Academic Press.

Schoenfeld, A. H. (1989). Teaching mathematical thinking and problem solving. In L. B. Resnick & L. E. Klopfer (Eds.), *Toward the thinking curriculum: Current cognitive research* (pp. 83–103). Alexandria, VA: Association for Supervision and Curriculum Development.

Schwab, J. J. (1960). What do scientists do? *Behavioral Science, 5,* 1–27.

Sharp, D., Bransford, J., Vye, N., Goldman, S., Kinzer, C., & Soraci, S. Jr. (1992). Literacy in an age of integrated media. In M. J. Dreher & W. Slater (Eds.), *Elementary school literacy: Critical issues* (pp. 183–210). Norwood, MA: Christopher-Gordon Publishers.

Silver, E. A. (1990). Contribution of research to practice: Applying findings, methods, and perspectives. In T. J. Cooney & C. R. Hirsch (Eds.), *Teaching and learning mathematics in the 1990s: 1990 Yearbook of the National Council of Teachers of Mathematics* (pp. 1–11). Reston, VA: National Council of Teachers of Mathematics.

Spiro, R. J., Feltovich, P. L., Jacobson, M. J., & Coulson, R. L. (1991). Cognitive flexibility, constructivism, and hypertext: Random access instruction for advanced knowledge acquisition in ill-structured domains. *Educational Technology, 31*(5), 24–33.

Van Haneghan, J. P., Barron, L., Young, M. F., Williams, S. M., Vye, N. J., & Bransford, J. D. (1992). The Jasper Series: An experiment with new ways to enhance mathematical thinking. In D. F. Halpern (Ed.), *Enhancing thinking skills in the sciences and mathematics* (pp. 15–38). Hillsdale, NJ: Erlbaum.

VanLehn, K., & Brown, J. S. (1980). Planning nets: A representation for formalizing analogies and semantic models of procedural skills. In R. E. Snow, P. Federico, & W. Montague

(Eds.). *Aptitude, learning and instruction* (Vol. 2 , pp. 95–137). Hillsdale, NJ: Erlbaum.

Vygotsky, L. S. (1978). *Mind in society.* Cambridge, MA: Harvard University Press.

Whitehead, A. N. (1929). *The aims of education.* New York: Macmillan.

Williams, S. M. (1992). *Putting case-based instruction into context: Examples from legal, business, and medical education. Journal of the Learning Sciences, 2,* 367–427.

Williams, S. M., Bransford, J. D., Vye, N. J., Goldman, S. R., & Carlson, K. (1992, April). *Positive and negative effects of specific knowledge on mathematical problem solving.* Paper presented at the annual meeting of the American Educational Research Association, San Francisco.

Wolf, D., Bixby, J., Glen, J., & Gardner, H. (1991). To use their minds well: Investigating new forms of student assessment. In G. Grant (Ed.), *Review of research in education* (Vol. 17, pp. 31–74). Washington, D.C: American Educational Research Association.

Yackel, E., Cobb, P., Wood, T., Wheatley, G., & Merkel, G. (1990). The importance of social interaction in children's construction of mathematical knowledge. In T. J. Cooney & C. R. Hirsch (Eds.), In *Teaching and learning mathematics in the 1990s: 1990 Yearbook of the National Council of Teachers of Mathematics* (pp. 12–21). Reston, VA: National Council of Teachers of Mathematics.

CHAPTER 5

For Research to Reform Education and Cognitive Science

James G. Greeno

M any people advocate fundamental reform of the American educational system, especially in mathematics and science. In this chapter, I focus on a general issue on which nearly all advocates of educational reform agree. Advocates object strongly to the receptive character of most school learning and propose that students should be more active as participants in the construction of their knowledge (e.g., California State Board of Education, 1992; Mathematical Sciences Education Board, 1990; National Council of Teachers of Mathematics, 1989).

This effort to reform educational practice has a companion in cognitive science, where some researchers are working to develop scientific practices, knowledge, and a theory that challenge those that are currently prevalent. In this alternative view, which may be called *situativity theory* (Greeno & Moore, 1993), cognition is viewed as interaction in which

Supported by National Science Foundation Grant MDR-9053605.

agents participate with other people and physical systems. In situativity theory, patterns of interaction that people can learn are considered practices, and learning is considered to be a process of becoming more able to participate in practices. The scientific development of this view is in an early stage, and it includes at least four different lines of work that are only weakly connected: *ecological psychology* (e.g, Gibson, 1979/1986; McCabe & Balzano, 1986; Turvey, 1992), *ethnographic anthropology and sociology* (e.g., Hutchins, 1991a; Lave & Wenger, 1991; Suchman, 1987), *situation theory* (e.g., Barwise, 1989; Barwise & Perry, 1983), and a form of educational research involving detailed study of *interactive activity in learning settings* (e.g., Cobb, Wood, & Yackel, 1990; Lampert, 1990).

Situativity theory suggests a perspective on the agenda of educational reform. This view focuses on the practices that students learn in school, with the contents and skills acquired in subject-matter disciplines playing an important but not exclusive role. Important issues include students' learning to formulate and to recognize significant questions and hypotheses, to construct and to evaluate arguments with evidence and examples, and to use concepts and principles as resources for understanding events and systems in their experience. From this perspective, the most common educational practices in our schools are severely flawed. The practices of cognition and learning that successful students learn are mainly to memorize in order to recite, and to rehearse in order to perform routine cognitive procedures. These are grotesquely inadequate cognitive practices from the perspective of any plausible version of what people need to be successful in their lives and work. Situativity theory suggests a strong form of the position that focuses on educational goals of learning to engage in intellective practices, especially social activities, that depend on concepts and principles of subject-matter disciplines (e.g., J. S. Brown, Collins, & Duguid, 1989).

It is widely believed, perhaps mistakenly, that the educational impact of research in cognitive psychology has been small. It seems to me that it has been pervasive and profound. Educational practice in the United States is shaped fundamentally by the view of cognition and learning that has dominated American psychological research, a view that focuses on individual knowers and learners who acquire knowledge and cognitive skills by adding small pieces incrementally to what they have learned previously. This view has been developed in great detail in the psychological research literature. It provides basic assumptions that underlie the organization of our school curriculum, assessments of student achievement, and important aspects of teachers' classroom practices, especially in the areas of mathematics and science.

Research can also play a significant role in supporting educational reform. This would involve developing a view of cognition and learning that differs from the individualistic and elementaristic perspective that we have focused on. Assumptions of situativity theory provide a promising framework for such a program of research.

Research conducted in educational settings can also play an important role in developing the fundamental knowledge, theory, and scientific practices of the study of cognition and learning as situated activity. A metatheoretical commitment of situativity theory is to treat the social organization of activity as integral to the processes of cognition and learning, rather than simply as a context that modulates individual cognitive and learning processes. As many social scientists have remarked, laboratory experiments have a social organization, albeit a specially constructed one. We may need to study cognition and learning in settings that are organized in more complicated ways so that the systems that we study can inform us about fundamental psychological principles. School learning provides one such opportunity.

This discussion considers research on three topics that provide significant information relevant to education practices in mathematics and science: (a) communities of practice, (b) conceptual understanding and growth, and (c) cognitive structures.

Communities of Practice

Until recently, most research on cognition and learning has studied either individuals or social organizations, and the two sets of results have had little influence on each other. Nearly all psychological research has considered cognition and learning as achievements by individuals, with social conditions considered as a part of the context. Anthropologists and sociologists have considered cognition and learning as social phenomena organized in group interactions. Considerable progress has been made in each of these perspectives, but our scientific situation has required us to focus either on individuals or on organized groups, with little or no meaningful connection between the understandings that have been achieved at the two levels.

Recent research is beginning to provide analyses in which concepts that refer to individual and social processes are interrelated. Examples include Hutchins's (1990, 1991b) studies of navigational teams on naval vessels and of cockpit crews of commercial airlines. Another example is Wenger's (1990) study of workers in an insurance claims office where expert-systems technology was introduced, changing the nature of their work.

Wenger (1990) developed an analytical concept, *communities of practice*, that is promising for integrating individual and social perspectives on cognition and that has become an organizing theme of research at the Institute for Research on Learning. The potential integrative power of the idea comes from the complementary nature of group interaction and the participation of individuals. It is natural, in this perspective, to attend to properties of a group's interactions and to the varieties of individuals' participation in the group activities.

Most studies of learning, reasoning, and understanding in cognitive science have treated these as processes that occur at the level of individual cognition and have attempted to discover their properties. These studies, considered as practices that occur in communities, assume that these processes are not fixed but are aspects of social practice and that they occur in different ways according to patterns of social interaction and of individual activity learned in communities.

Lave and Wenger (1991) contributed a seminal discussion of learning from the perspective of situativity theory. They discussed research on learning in communities of practice, including traditional birth attendants in Mexico (Jordan, 1989), apprentice tailors in Liberia (Lave, 1993), quartermasters in the U.S. Navy (Hutchins, in press), apprentice butchers in the United States (Marshall, 1972), and members of Alcoholics Anonymous (Cain, 1991). They developed an important concept of *legitimate peripheral participation,* by which they refer to ways in which newcomers in a community can learn to participate in the practices of the community as participants with less central responsibilities for the community's overall success than they will have in the future.

Four lines of research on communities of practice seem to be particularly promising. They include a kind of study that John Seely Brown (1991) called *pioneering research,* studies of reasoning in everyday practices, studies of school work as everyday practice, and studies of personal epistemologies.

Examples of Pioneering Research

J. S. Brown (1991) proposed the term *pioneering research* for a kind of research that does not fit into either of the standard categories of basic or applied research. Pioneering research is the study of practices in real-life situations in which people agree that there should be a change in their activities. It is fundamental as well as practical research because it examines and analyzes basic assumptions that underlie the practices that occur in the situation. A goal is to understand both the assumptions that support the practices that people want to change and those that support desirable changes in the practices. Researchers and practitioners collaborate in the research with complementary and mutually supportive goals. The researchers' primary goal is to use observations of the practices to infer general principles of social and individual activity. The practitioners' primary goal is to make their activities more effective, and a better understanding of the organization of those activities can facilitate their improvement. It can be particularly valuable to study practices in an organization that is undergoing change, either because the practitioners are deliberately reforming their practices or because of a change in the circumstances of their activities.

Pioneering research about educational practices involves working in a setting of educational practice to understand and to change the nature of practices in that setting, as well as to articulate assumptions that underlie the practices. A classic example in educational research was a study by Harold Fawcett (1938), who developed a course in high school geometry that focused on students' acquiring practices of mathematical discourse, including formulation of definitions and postulates and eval-uation of arguments. Each student constructed her or his own version of geometry, including the definitions and postulates that he or she chose to formulate and the theorems that he or she chose to derive. A significant part of the course involved applying principles of deductive reasoning to everyday issues, as well as to the mathematical topics of geometry. Faw-cett's discussion of the course presented his assumption that to learn the practices of mathematical reasoning, including carefully formulating def-initions and requiring explicit statements of assumptions, students need to engage in those practices. It is not sufficient to have them displayed in texts and illustrated in demonstrations given by a teacher.

A contemporary example along the same lines is research that Mag-dalene Lampert (1990) conducted by teaching a fifth-grade mathematics class. Lampert's students engaged in discourse about mathematical con-cepts and representations, offering conjectures and providing reasons for their hypotheses, and recognizing occasions in which they revised their thinking. Part of Lampert's research effort is to articulate assumptions about teaching, learning, and norms of discourse that are critical for the success of her teaching. One of her assumptions is that productive math-ematical discourse requires that participants observe norms of modesty and courage, involving willingness to communicate their opinions and to consider the opinions of others, treating all opinions (including theirs) as being open to revision, and requiring adequate reasons as conditions for changing their opinions.

Further examples include projects focusing on the middle-school level directed by Ann Brown, Joseph Campione, and their associates (A. Brown, 1992) in which they collaborate with the administrators and teach-ers of a school to conduct a form of education that they call a *community of learners*, and a project directed by Marlene Scardamalia and Carl Ber-eiter (1991) in which students actively communicate with each other using a computer-based network. Students are engaged primarily in research activities, which in Brown and Campione's project focus on biological themes such as selective adaptation and extinction, and in Scardamalia and Bereiter's project include a variety of topics in natural science, social science, and the humanities. These studies examine the assumption that students should learn to engage in reading, writing, and other representational work that is instrumental to their participation in a group's project activities, thus developing their skills in language and communication in contexts of broader projects.

Another example is the Jasper Project of John Bransford and his colleagues (the Cognition and Technology Group at Vanderbilt, 1990; see also chapter 4, this volume). An assumption of traditional mathematics education is that, by solving word problems, students learn to apply mathematical procedures in everyday situations. Bransford and his colleagues assume that solving textbook problems often results in knowledge that is inert in many situations in which mathematical thinking could be useful. The Jasper Project has developed materials for mathematical problem solving on the basis of another assumption: Learning to use mathematics occurs when students understand a situation with quantitative properties and reason about those quantities in relation to significant questions about the situation. Students see videotapes that portray meaningful situations in which quantitative questions matter (e.g., Will there be enough time to accomplish a needed trip), determine what information is needed to make inferences, and search the tape to obtain the information.

Two more examples, involving younger children, are studies conducted by Cobb et al. (1990) and by Resnick, Bill, and Lesgold (in press). These investigators assume that children construct their mathematical knowledge in discourse activities. They work with teachers who present mathematical problems that children work on in classroom discussions or in small groups, and children's progress in learning mathematics is inferred from the explanations that they give each other in the course of their work.

Everyday Practices

A growing body of research is providing insight into reasoning that occurs in everyday practices, especially those involving quantitative reasoning in commercial transactions and other situations. An impressive variety of quantitative reasoning was described by Terezina and David Carraher and Annalucia Schliemann (Carraher, Carraher, & Schliemann, 1985) and by Geoffrey Saxe (1990), who observed street vendors, carpenters, and fish sellers. Jean Lave and her students (Lave, 1988; Lave, Murtaugh, & de la Rocha, 1984) observed people shopping for groceries and managing their diets. Sylvia Scribner and her associates (Martin & Scribner, 1988; Scribner, 1984) analyzed reasoning in the work settings of a dairy warehouse and of a machine-tool shop. Edwin Hutchins (1990, 1991b) analyzed performance in the work settings of a navigation room on a ship and of the cockpits of commercial airplanes.

Another growing line of research is the ethnographic study of practices in scientific laboratories (e.g., Latour & Woolgar, 1979; Lynch & Woolgar, 1990). These analyses provide information about important everyday problem solving that is crucial in the conduct of research. This research about the practices of scientific research reminds us that intellective activity always has a practical, everyday aspect and that, on many occasions, its reflective or analytical aspects are no more visible than they are in everyday practices that occur in contexts such as shopping, selling merchandise, or navigating a ship.

School as a Setting of Everyday Practices

Lave, Smith, and Butler (1988) pointed out that we need to understand children's activities in school as everyday practices with their own integral organization. Everyday practices in classrooms include the routines that students and teachers follow in presentations of material by teachers, recitation by students, production of seatwork and homework papers, and grading of students' papers. Everyday school practices also include the patterns of social interaction that teachers and students participate in, usually involving the teacher presenting information and asking students questions to determine whether they understand, and students displaying that they have or have not assimilated the information that the teacher has presented. Lave et al. pointed out that these practices are usually focused more on resolving dilemmas of performance on instructional tasks than they are on dilemmas of the subject matter.

Penelope Eckert (1989) contributed an important analysis of the social organization of a high school in which most students viewed themselves and each other as members of one or the other of two broad social groups that called themselves "jocks" and "burnouts." Most jocks were from middle-class families, they participated actively in the official extracurricular activities of the school, and they viewed school as providing them with valuable preparation for functioning successfully in the social life of a college, where they expected to be in the next significant stage of their lives. Most burnouts, on the other hand, were from working-class families, they had significant personal investment in not participating in official school activities, and they viewed school as impeding their preparation for the industrial work, family lives, and participation in the working-class communities where they expected to be in the next significant stage of their lives. In the students' academic work, these different orientations were especially evident in their participation in mathematics and science classes. Burnouts rarely took any mathematics or science courses that were not required for satisfying minimum requirements, viewing courses in mathematics and science as typical examples of the hierarchical and authoritarian structure of the school that they rejected (Eckert, 1990).

Personal Epistemologies

Another set of important issues about active learning are addressed in research about different ways in which individuals understand the nature of knowledge and learning and understand themselves as cognitive and learning agents. Examples of studies of individual epistemological differences include contributions by Belenky, Clinchy, Goldberger, and Tarule (1986) and by Dweck and Legett (1988), which have shown that individuals differ significantly in the degree to which they understand learning as a participatory process in which they play active roles as agents.

Viewed in the perspective of communities of practice, individual epistemological differences raise compelling questions both for scientific understanding and for practical reform. It seems unlikely that an individual's beliefs and approaches to learning are uniform in all the social situations that he or she participates in. Eckert's (1989) results suggest that different individuals' beliefs about learning may involve tacit assumptions about the activities that count as "learning" in the social environment. Jocks and burnouts treated information differently: For jocks, information was often treated as a kind of possession that was used to maintain or enhance one's status, whereas for burnouts, information was more likely to be treated as a shared resource. Students' views of themselves as learners undoubtedly interact with their views of knowledge and information in social contexts. The nature of these interactions need to be investigated, and we need to take them into account in work on reforming the social organization of school learning.

Contributions to Scientific and Educational Progress

Studies of educational processes focusing on communities of practice have considerable promise for expanding and for reformulating the psychology of learning, and for supporting fundamental educational reforms. The studies of cognition in communities of practice begin to provide a basis for understanding processes of learning and cognition that would expand and reformulate the psychology of learning considerably. They also provide evidence that, so long as we understand cognition and learning only as individual processes, we will fail to grasp aspects that are crucial to bringing about effective change. The changes that are needed include changes in the social organization of schooling, involving fundamental changes in the roles of students and teachers in the construction of knowledge, which can be addressed in research that uses concepts and methods that are drawn from traditions of the cognitive sciences as well as of the social sciences (Eckert & Knudsen, 1992).

We face a challenge in the study of communities of practice in education, and in integrating the insights that are developing in studies of group interaction with understanding of cognition and learning involving the contents of subject-matter disciplines. Some of the issues that we need to incorporate in this study are addressed in studies of conceptual understanding and conceptual growth.

Conceptual Understanding and Conceptual Growth

An important proposal for educational reform is that students should be more active participants in classroom discourse. They should contribute their understandings of concepts and participate in the construction of explanations and meanings. This is feasible only if students have understandings to contribute. A traditional view is that students lack understanding, especially in mathematics and science, until they have been taught the correct meanings of concepts.

An alternative view is that students bring rich intuitive understandings to their learning of mathematical and scientific concepts and that their learning consists of shaping, refining, and learning to apply their understandings rather than of acquiring representations of concepts by storing information structures in memory. Their intuitive understandings constitute a valuable resource for conversations in which they can participate, and for conversations about their own and other students' understandings are a crucial part of the learning process.

The question of whether students have understandings that enable them to participate productively in discourse that fosters growth of conceptual understanding in communities of practice is a scientific question, and researchers are addressing it in different ways. One line of research focuses on the informal understanding that students have about systems and processes, especially in the domains of mathematics and science, that should be taken into account as we design learning activities for them. Another is analysis of activities in which students may engage from the point of view of concepts in subject-matter domains such as mathematics and science.

Development of Conceptual Understanding

Research investigating growth of children's understanding of concepts is a rich and growing enterprise in developmental psychology. The problem was set by Piaget (e.g., Inhelder & Piaget, 1959/1964), who found general patterns in children's reasoning abilities that he interpreted as growth of general schematic operational structures. More recent research has shown that children have significant conceptual understanding and reasoning abilities at earlier ages than appeared to be the case in Piaget's studies and has identified patterns of conceptual growth that can be characterized as the development of informal theories in which children's understanding becomes more differentiated and is restructured as they grow older and more experienced in reasoning and communication in their society. Examples of this developmental research include studies of children's understanding in the domains of biology (Carey, 1985; Hatano & Inagaki, 1987), psychology (Wellman, 1990), number and quantity (Gelman & Gallistel, 1978), and physics (diSessa, in press).

Practices and Subject-Matter Content

The innovations that are advocated as reforms in mathematics and science education face an important challenge because mathematics and science education currently emphasize covering a large body of subject-matter content. In traditional instruction, each lesson is designed to cover a topic, and the topics are related to each other systematically, forming a content curriculum. Textbooks and teachers are held accountable for covering those parts of the curriculum that are considered essential and are given extra credit when they cover some of the optional topics.

When instruction is organized around practices of mathematical or scientific reasoning and discourse, the relations of classroom activities to a content curriculum become less clear. Discussions are formed around problems and issues, and many different mathematical or scientific topics can come up. For example, a discussion about the distance that a ship travels in 3 1/2 hours at a speed of 40 knots included the standard topics of rate, multiplication, fractions, and the number line (because students drew diagrams of a line with marks for the distances after each hour; Magdalene Lampert, personal communication, March 1992).

The relation of subject-matter topics to classroom activities involves a theoretical question about concepts and principles. A person's understanding of a concept or principle has both *explicit* and an *implicit* aspects, which are evaluated with somewhat different kinds of evidence. Explicit understanding of a concept or principle is shown in discourse about the concept, for example, by defining it, explaining its meaning, and discussing its application and its implications in situations. Implicit understanding of a concept or principle is shown when a person acts in ways that are consistent with the concept or principle, thereby performing some task correctly, where correct performance is interpreted as an example of the concept or principle in use.

Development of a theory of implicit understanding is a significant goal of cognitive science research. It could also play a significant role in the effort to reform educational practice, especially in mathematics and science. Its role in practice could be in providing a systematic way of making classroom activities accountable to the subject-matter curriculum. This will occur if the theory supports analyses of classroom activities that show how performance of those activities is evidence for students' implicit understanding of concepts and principles that constitute the content curriculum.

I have been working on a version of such a theory, along with Joyce Moore, Rory Mather, Randi Engle, and others. The example that we are studying is implicit understanding of the mathematical concepts of variable and linear function. The material for this research is the performance by middle-school and high school students on tasks that use the apparatus shown in Figure 1.

We call the apparatus a *winch system*. It is a board about one yard long with two tracks, each marked with distances in inches. Each track has a small metal block, attached by a string to a spool that is turned by a handle. When the handle is turned, the block is pulled along the track. Spools vary in size. The axles can be connected so that both spools rotate when either handle is turned, or they can be disconnected so that the handles work independently. An apparatus like this was used by Piaget, Grize, Szeminska, and Bang (1968/1977) in their studies of children's understanding of quantitative functions, although in their apparatus numerical scales were not provided.

In one study (Greeno, Moore & Mather, 1992), we asked students to solve problems (e.g., Can you make both blocks get to 24 at the same

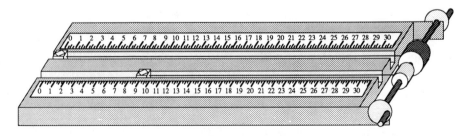

FIGURE 1. Winch apparatus used in research on implicit understanding of functions.

time?) and answer questions about what would happen in a specific situation (e.g., If the blue block has a six-spool and starts at 0, and the red block as a three-spool and starts at 9, and we turn the handles together, will the blue block catch up to the red block? When will that happen?). We asked teachers to recommend students who would probably score near the national average on mathematics achievement tests. We interviewed two students at a time, with the students in each pair being friends and of the same sex. Some of the students were in 7th grade and had not studied algebra, some were in 9th grade near the end of their first-year course in algebra, and some were in 11th grade near the end of their second-year course in algebra.

Students generally solved most of the problems and answered most of the questions correctly. Their success indicates that they had significant implicit understanding of the concepts of variable and linear functions. We developed an analysis of the performance that we observed and of the concepts of variable and linear functions to show how the concepts are involved in performance (Greeno et al., 1992).

The key ideas in our analysis are the idea of *constraints* in situation theory (Barwise, 1989; Barwise & Perry, 1983; Devlin, 1991) and the idea of *affordances* in ecological psychology (e.g., Gibson, 1979/1986). This idea of constraints in situation theory is very general; it includes any regularity in situations and activities. Gibson's idea of affordances refers to ways in which systems in the environment support activities in which people interact with those systems. Successful activity by people depends on their being attuned to constraints and affordances.

People who engage in an activity communicate about it for several reasons, including coordinating what they do, agreeing on goals and plans, and helping novices to learn to be more successful in the activity. We suppose that concepts arise in communicative interactions about activity. The things that people find important to discuss generally involve constraints and affordances that are significant in what they are doing. We suppose, therefore, that concepts usually refer to constraints and affordances. The specific constraint or affordance that a term refers to varies from one situation to another, although there is considerable overlap in the properties and relations that are involved. It seems most appropriate to call the meaning of a concept a family of constraints and affordances.

Concepts have at least three roles in social activity. Most directly, they function in discourse that is instrumental in activities, which is how we think that they originate. Terms are used to refer to aspects of activity in people's planning, setting goals, and otherwise coordinating what they do.

A second function of concepts is as objects of theoretical inquiry. Meanings of concepts are made explicit and are used to formulate propositions that are examined in relation to other concepts and propositions, the process that Dewey (1938) called *reflective inquiry*. A collection of such propositions, taken together, constitutes a theory that may be stated in formal terms.

A third function of concepts, dependent on the second, is in theoretical discourse about activity. A community develops conventions of discourse about significant topics. Some of the topics are the activities that are important in the communities, that is, the community develops ways of talking about its activities, as well as ways of talking in its activities. For example, assessment of individuals and groups is one important function of discourse about activities. When an individual or a group is assessed, their performance is evaluated in terms of some concepts that the community uses to refer to significant aspects of activity. Such assessments are theoretical, in the broad sense used here, in that they provide analyses of performance in terms of concepts that are used (more or less) systematically in describing such performances in the community.

Performance may be assessed in terms of concepts that the performer knows, but it may also be assessed in terms of concepts that the performer does not know explicitly. A familiar example involves young children's language behavior. We can easily assess whether a sentence uttered by a 3-year-old agrees with a rule of verb–subject agreement in number. The rule of verb–subject agreement is a constraint on word usage. An assessor's knowledge may include knowing the term *verb–subject agreement* that refers to that constraint, and the assessor may use that term in describing what the child says. The child's language behavior may or may not comply with the constraint, independently of whether the child can talk about the linguistic constraint. In most studies of children's language behavior, interviewers do not even bother to ask children to explain why they use words in the way that they do.

We consider understanding of a concept or principle as attunement to constraints and affordances that constitute that concept or principle. *Explicit understanding* includes attunement to constraints and affordances of representing the constraint, most commonly in language. *Implicit understanding* of a concept or principle involves attunement to constraints and affordances in activity in which the constraints and affordances have significant functions, but the activity does not include representing that concept or principle, that is, in acting systematically in accord with the constraints and affordances, rather than talking about or otherwise representing the constraints and affordances.[1]

What the constraints and affordances that constitute a concept or principle are has to be grounded in some community's practice. We suppose that the articulation of concepts and principles in explicit language occurs in theoretical inquiry by a community that may be of recognized specialists in the activity domain. This fits the pattern of mathematical concepts well. Mathematicians are responsible for developing accounts of concepts and are consulted about the meanings of those concepts in many contexts, particularly in educational activities.

Regarding the situation that we use in our research on linear functions, there are concepts of *variable* and of *linear functions* found in the practice of mathematics, or, more correctly, in the practice of high school mathematics education. The problems and questions that our student participants answered were about quantitative relations that are examples of variables and of linear functions. We are the ones, however, who know the concepts of variable and linear functions explicitly. We participate in

---

[1]Talk about a constraint may include a term that refers to the constraint, such as verb–subject agreement, but a single term need not be used. For example, if a child correctly says, "My doggie chases cats," an interviewer might ask, "Could you say, 'My doggie chase cats'?" and the child might answer, "No, there's just one." This would represent the situation type involved in the constraint, although not with the same sophistication that someone with more theoretical linguistic training might have.

discourse practices that include the terms *variable* and *linear functions* and use them in conversations about quantities.[2] Our research goals include developing an analysis of these concepts and of the activities of the student participants that shows how performance on the problems and questions can be used to infer that the students were attuned to constituent constraints and to affordances of the mathematical concepts.

Attunement to Ordinal Constraints

Consider the problems that we asked the students to solve, such as, "How can you make the blue block get to 20 ahead of the red block?" and "Can you make both blocks get to 24 at the same time?" Students gave two kinds of solutions, which we call *ordinal* and *metric*. Ordinal solutions specified an ordinal relation among variables (e.g, "You can make the spool bigger over there than over here, and make the same number of turns for each person"). Metric solutions specified numerical values for the relevant variables. I consider the ordinal solutions here. (I discuss reasoning about specific numerical values of other problems shortly.) For the first problem, nearly all 9th- and 11th-grade pairs gave correct solutions immediately, and 11 of the 13 solutions given by 7th-grade pairs were either immediately correct or were correct after some discussion. Every student pair solved the problem initially by saying that both tracks should have the same spool size and starting position and that the handles should be turned together. We then asked whether it could be done another way or, in some cases, specified that one of the variables (e.g., the spool size) should differ between the tracks. Every pair gave at least one solution involving compensating variables, and the 9th- and 11th-grade pairs gave an average of two such solutions.

Ordinal solutions provided evidence for students' implicit understanding of the concept of *variable*. For these problems, the students

---

[2]Our student participants' understanding included abilities to represent many of the constraints that constitute the concepts of variable and linear functions, such as the fact that a block moves at a constant distance per turn and a larger spool makes its block move farther on each turn. They did not, however, engage in the kind of explicit representation referred to here as theoretical inquiry. That is, they did not discuss the meanings of the concepts of variable and linear functions explicitly to any significant extent.

needed to understand covariation among variables. There are some basic ordinal constraints in the physical system that the students understood, including the constraint that turning the handle causes a change in the position of the block connected to the handle, and that turning the handle more increases the distance that the block is moved. More complicated constraints are involved in the reasoning about relations between positions of the blocks, and the students understood many of these as well. For example, if both blocks start at the same position and the handles are turned together, but the spool sizes differ, then the block with the larger spool will be ahead. Or, if the spool sizes are the same, but one handle is turned more than the other, the block with more turns will move farther.

A compact representation of ordinal constraints may be given using confluence equations (DeKleer & Brown, 1984), which have the form $\Delta q1 \pm \Delta q2 \pm \ldots \pm \Delta qn = 0$. Each variable can take values $+$, $-$, or 0. A solution of the equation is a set of values that is consistent with the equation. A set of confluence equations for the winch is as follows:

$$\Delta q\text{spoolsize} - \Delta q\text{distperturn} = 0, \tag{1}$$

$$\Delta q\text{disperturn} + \Delta q\text{turns} - \Delta q\text{distance} = 0, \tag{2}$$

$$\Delta q\text{startposition} + \Delta q\text{distance} - \Delta q\text{endposition} = 0. \tag{3}$$

In Equation 1, Δqspoolsize is the difference between the sizes of the spools in the two tracks, and Δqdistperturn is the difference between the distances that the blocks move on a single turn. Because the spool size determines the distance that a block moves on each turn, a positive (or negative, or zero) difference in one of these variables goes with a positive (or negative, or zero) difference in the other. This corresponds to the solutions of Equation 1, which is satisfied if the two variables have values $(+,+)$, $(-,-)$, or $(0,0)$.

In Equation 2, Δqturns is the difference between the number of turns of the two handles, and Δqdistance is the difference between the total distances that the two blocks travel. If one of the three variables is 0, the other two are constrained. For example, if Δqdistperturn $= 0$, then (Δqturns, Δqdistance) must be $(+,+)$, $(-,-)$, or $(0,0)$. This represents the fact that if the distances per turn are equal (i.e., the spool sizes are the same), then more turns of one handle cause that handle's block to move a greater distance. Another example is that if Δqdistance $= 0$, then (Δqdistperturn, Δqturns) must be $(+,-)$, $(0,0)$, or $(-,+)$. This represents the fact that if the two blocks move the same total distance, then either they have the same distances per turn and numbers of turns, or one track has more turns and the other track has a greater distance per turn.

In Equation 3, Δqstartposition and Δqendposition are the differences between the starting positions and the ending positions of the blocks. For example, for the ending positions to be equal, Δqendposition $= 0$, either the starting positions and distances are both equal (Δqstartposition, Δqdistance) $= (0,0)$, or the starting position of one track is ahead of the other, and the distance that the other block moves is greater (Δqstartposition, Δqdistance) $= (+, -)$ or $(-, +)$. As another example, to have Δqendposition $= +$, we can have either Δqstartposition $= +$, or Δqdistance $= +$, or both. If Δqstartposition and Δqdistance are $(+, -)$ or $(-, +)$, then the value of Δqendposition is indeterminate, that is, Δqendposition could be $+$, $-$, or 0, depending on the numerical values of Δqstartposition and Δqdistance.

The ordinal solutions that students gave to problems such as "How can you make the blue block get to 20 ahead of the red block?" and "Can you make both blocks get to 24 at the same time?" are evidence that they were attuned to constraints of the kind represented by Equations 1, 2, and 3. For example, getting the blue block to 20 ahead of the red block means that Δqendposition $= +$ (with the direction of Δ being blue minus red). One solution of Equation 3 has Δqstartposition $= +$ and Δqdistance $= 0$, which is a satisfactory solution. Another solution has Δqstartposition $= 0$ and Δqdistance $= +$. To have Δqdistance $= +$, in Equation 2, we can have either Δqdistperturn $= +$ and Δqturns $= 0$ or Δqdistperturn $= 0$ and Δqturns $= +$; Δqturns $= +$, with Δqstartposition $= 0$ and Δqdistperturn $= 0$, gives a satisfactory solution. To have Δqdistperturn $= +$, in Equation 1, we need Δqspoolsize $= +$.

Our assumption that students were attuned to constraints that can be expressed as Equations 1, 2, and 3 does not imply that they had representations of Equations 1, 2, and 3 stored in their memories. Rather, we hypothesize that students' reasoning was attuned to ordinal constraints because they had abilities to construct and to enact mental simulations that behaved according to those constraints. This hypothesis is consistent with other discussions of mental models in reasoning, such as those by Johnson-Laird (1983) and Yates et al. (1988). We hypothesize that students reasoned by using mental simulations of events of moving the blocks by turning a handle or the handles under different conditions and that their attunement to constraints (e.g., Equations 1, 2, and 3) was in the ways in which these simulations worked. For example, rather than assuming that a symbolic rule, such as "If larger spool size, then greater distance per turn," was stored in memory, we assume that students could simulate turning a handle with different spool sizes and linked axles and that their enactment included simulating the block with the larger spool moving farther on each turn and thereby getting farther ahead with each simulated turn.

We hypothesize that when students gave ordinal solutions for problems, they enacted simulations with properties corresponding to the conditions given in the problem. For example, to solve "How can you make the blue block get to 20 ahead of the red block?", students could enact a simulation of the blocks moving with equal starting positions and spool sizes with the handles linked, and register that the blocks would always have equal positions, and then transform their simulation so that the simulated spool size for the blue block was larger than the simulated red block's spool size, and enact a simulation in which the blue block was ahead of the red block increasingly as they moved along the track.

Attunement to Metric Constraints

Now consider questions that required numerical answers, such as "If the blue block has a six-spool and starts at 0, and the red block has a three-spool and starts at 9, and we turn the handles together, will the blue block catch up to the red block? When will that happen?" We hypothesize that students enacted mental simulations to answer the first part of the question. The students and the interviewer set up the winch apparatus so that the premises of the question were true. Therefore, in answering "Will the blue block catch up to the red block?", students looked at the apparatus with the larger spool connected to the blue block and the red block located ahead of the blue block. We hypothesize that they simulated the motion of the blocks mentally when a handle was turned and that their simulations had a property corresponding to the blue block moving farther on each turn and overtaking the red block within a few simulated turns.

To answer "When will that happen?", however, we hypothesize that another process was needed. Our hypothesis about students' mental simulations is that they support inferences about ordinal properties of events but not metric properties.

Students in our experiment used four distinguishable methods in answering questions of this kind. One method used mental arithmetic. A student gave an answer (e.g., "At the third turn, they'll both be at 18"), and if the interviewer asked, "How did you figure that out?", the student reported some arithmetic calculations. One student said, "I just kept on adding 3—3 and 6." In another pair, a student said, "After four turns. After three turns, it'll be tied and after four turns it'll be ahead." Answering, "How did you figure that?", the students said, "I just multiplied in my head. . . . 6 and 4 is 24, uh, 6 and 3 is 18, 3 and 3 is 9 and add that 9 already, it's 18 and 18, it'll be tied and then so they'll equal, so they just take one more turn."

Another method used by students involved writing a table of numbers, starting with the initial positions of the blocks and adding 3 or 6 successively until the numerals in the two columns were equal.

In a third method, students pointed to positions along the two tracks, moving ahead by 3 inches along the track with the red block and moving ahead by 6 inches along the track with the blue block, counting the increments, until the positions they pointed to on the two tracks were equal.

A fourth method involved turning the handle, counting the turns, until the blocks were at the same position on the two tracks. This method was used in a few cases, when students said that they did not know how to figure out the answer to a question, and the interviewer invited them to try and see. In several cases, this method of direct observation was used by students to check whether an answer that they had given was correct.

These correct performances provide evidence for students' understanding of relations among quantities that we characterize as linear functions. The standard way of expressing quantitative relations is in the form of equations, for example,

$$y = b + a \times x, \tag{4}$$

where x is the number of turns, a is the size of the spool, b is the starting position, and y is the ending position.

It is useful for our analysis to have more differentiated statements of constraints. One distinction is between quantities and numbers. Arithmetic operations, such as addition and multiplication, are relations among numbers. In our terms, arithmetic operations are constraints in the domain of numbers. Quantities are properties of physical objects or events. Quantities and numbers are related, of course. Technically, the relations are measure functions that map physical objects or events to their numerical values. One distinction between quantities and numbers is that quantities have units, so, for example, 3 inches is a quantity, which has the number 3 as a component. A position along the track has the quantitative property of being 3 inches from the origin of the ruler, an event of moving a block can have the quantitative property that its distance is 3 inches, and a spool can have the quantitative property that its circumference is 3 inches. In all of these cases, the number 3 is the measure of the quantity. Results of numerical operations on numbers, as measures of quantities, can be measures of other quantities.

Students who used arithmetic operations to infer answers to questions showed attunement to the constraints of arithmetic, as well as constraints that relate arithmetic operations to quantitative relations in the physical system. The relevant constraints may be stated as follows:

$$n\text{endposition} = n\text{startposition} + n\text{distance}, \tag{5}$$

$$n\text{turns} = r \Leftrightarrow n\text{distance} = (n\text{spoolsize}) \times r = (n\text{distperturn}) \times r. \tag{6}$$

This notation uses variables starting with n to denote numbers that are the measures of quantities, for example, if there are three turns, then $n\text{turns} = 3$. Constraints 5 and 6 are equivalent to the formula of linear

Equation 4, when the variables of 4 are interpreted as the measures of quantitative properties of the winch.

Students who used arithmetic operations of multiplication and addition gave evidence that they were attuned to Constraints 5 and 6. For the example problem of inferring when the blue block will catch up with the red block, we hypothesize that one process of reasoning involved enacting a mental simulation of an event of moving the blocks by turning a handle in which the blue block caught up with the red block.

Another constraint is that every time the handle is turned, the position of its block increases by a constant distance, and this corresponds to adding a constant number to the measure of the previous position. Formally,

$$n\text{turns} = 1 \Leftrightarrow n\text{position} = n\text{prevposition} + n\text{distperturn}$$

$$= n\text{prevposition} + n\text{spoolsize}. \quad (7)$$

Constraint 7 applies to a single action of turning the handle, rather than to a set of r turns. Mathematically, Equation 7 implies Constraints 5 and 6 because of the mathematical relationship between addition and multiplication. Student performance that was attuned to Constraint 7, however, need not have been attuned to Constraint 6.

Consider students who constructed tables of numerals. They wrote the initial positions of the blocks, 0 and 9 in the example that we are considering. Then they added 6 in one column, wrote 6, and added 3 in the other column, and wrote 12. Then they added 6 and 3 again, and wrote 12 and 15. Finally, they added 6 and 3 again and wrote 18 and 18. These successive additions of numbers are consistent with Constraint 7. They have a result that is consistent with Constraints 5 and 6, but the process of repeated addition does not require attunement to the constraint that includes multiplication. Similarly, students who answered the question using repeated mental addition were attuned to Constraint 7, but not necessarily to Constraint 6. We hypothesize that a process of reasoning that included using a table or mental arithmetic also included enacting a mental simulation in which simulated movement of the blocks corresponded to writing or thinking the numerals that represented their positions.

JAMES G. GREENO

Another constraint involves quantities, rather than numbers. Each time the handle is turned, the block moves a constant distance along the track:

$$q\text{turns} = 1 \text{ turn} \Leftrightarrow q\text{position} = q\text{prevposition} \oplus q\text{distperturn}$$
$$= q\text{prevposition} \oplus q\text{spoolsize}. \quad (8)$$

The variables in Constraint 8 refer to quantitative properties of the physical system, rather than to numbers. The operation symbol $\oplus$ denotes a physical operation of combining quantities, a concatenation of collinear physical segments, so that the combined segment has length equal to the combined lengths. Constraint 8 implies 7 because the number that is the measure of the length of the combined segment is the sum of the measures of the lengths of the two combined segments.

Consider students who answered questions by pointing to positions along the ruler, marking intervals equal to the distances that a block would move on each turn. This performance gave evidence of attunement to Constraint 8, but not necessarily of attunement to 7 or 6. We hypothesize that this process included enacting simulations of the blocks moving along the track on each turn, counting 1-inch marks along the ruler (or possibly adding numbers, which would use Constraint 7) to determine where the blocks would be after the turn, and pointing to the positions that resulted.

To summarize our analysis of numerical reasoning about the winch, first, we analyzed the concept of linear function as it applies to the activities in the study. Constituents of the concept, according to our hypothesis, are expressed as Constraints 5, 6, 7, and 8. Relations between these constraints and some of the methods used by the students are shown in Table 1. Correct performance accords with all of the constraints, but depending on the method used, some of the constraints are satisfied because of the properties of the physical winch system or of the system of numbers rather than of the properties of what the students did. If the answer was observed after turning the handle, all of the constraints were satisfied because of the way in which the winch system works. If the students marked intervals (in the version in which intervals are counted, rather than added numerically), the students were responsible for marking intervals that were the correct length, but the additive and multiplicative

TABLE 1
Sources of Accordance With Constraints

Method	Agent	Physical and numerical system
Turning the handle	—	(5), (6), (7), (8)
Marking intervals	(8)	(5), (6), (7)
Table	(7), (8)	(5), (6)
Arithmetic, including multiplication	(5), (6), (7), (8)	—

relations of the numbers were produced by the winch. If the students used a table, their actions were responsible for agreement with the constraint of adding equal numbers, but the answer also agreed with the constraint of multiplication because of the way in which multiplication and addition are related. If the students used the arithmetic operations of multiplication and addition of numbers, then the attunement to all of these constraints was caused by the students' actions.

Contributions to Educational and Scientific Progress

A scientific theory, including analytical concepts and methods for the study of implicit understanding of subject-matter concepts, can play a key role in allowing analyses that can relate educational activities to the concepts needed in the curriculum. This theory may also be helpful in the scientific study of children's conceptual growth by allowing more explicit analyses of the conceptual contents of activities that children can accomplish as their conceptual understanding becomes more sophisticated.

At the same time, a theory of concepts constituted by constraints and affordances of activity can provide a significantly broader scientific treatment of the psychology of concepts than we have had in cognitive psychology, in which most discussions of concepts have been limited to activities of classification. Classification is one form of activity in which concepts function, but it seems desirable to work toward an account of conceptual understanding that can apply more broadly, and an account in terms of constraints and affordances seems promising as a way to proceed.

Cognitive Structures

The final topic of this chapter is research on cognitive structures, a topic that has been developed productively in cognitive science during its 3 1/2 decades of development and that has included substantial and increasing attention to educational materials since the mid-1970s (e.g., Klahr, 1976). In less than 20 years, a scientific capability has been developed for analyzing cognitive structures and processes that students need in order to succeed in routine instructional tasks. Along with a very strong development of cognitive analyses of reading and substantial developments in the psychology of writing, cognitive analyses of problem solving in school mathematics and science, particularly physics, provide proofs of concept for the wide applicability of information-processing cognitive models in the curriculum.

Analyses of cognitive structures and processes bring together two bodies of study and research in education. During the remarkable growth of academic psychology in America following World War II, learning and behavior were studied abstractly, without attention to the content of information that was learned. At the same time, efforts to change education, especially those focused on mathematics and science education in the 1950s and 1960s, concentrated on the subject-matter contents of the curriculum and on ways in which the structure of mathematical and scientific knowledge could be conveyed to students. Figure 2 depicts the situation as it was in the early 1970s, with a large body of knowledge and research practice in experimental psychology focused on processes of learning and behavior, with little attention to content, and a large body of analysis of the organization of subject-matter content for curricula, with little attention to processes of learning and behavior.

The development of information-processing cognitive science that occurred by the early 1970s allowed analyses that addressed both the processes of learning and behavior and the contents of information that are learned and used in tasks of solving problems and answering questions. Many analyses are now available that do that. Examples include analyses of calculation procedures for elementary arithmetic (e.g., J. S. Brown & Burton, 1978; Burton, 1982), analyses of processes for understanding and solving word arithmetic problems (e.g., Briars & Larkin, 1984;

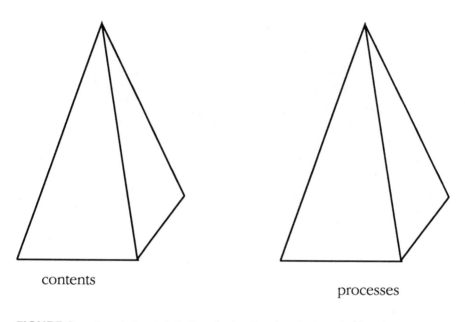

contents

processes

FIGURE 2. Knowledge and studies of educational contents and of learning processes
have been separate, until recently.

Kintsch & Greeno, 1985; Riley & Greeno, 1988), analyses of processes for
solving proof exercises in high school geometry (e.g., Greeno, 1978), and
analyses of processes for understanding and solving text problems in
physics (e.g., Chi, Feltovich, & Glaser, 1981; Larkin, McDermott, Simon,
& Simon, 1979; Simon & Simon, 1978). Some attention has been given to
processes of learning, including models of learning to solve problems in
arithmetic (VanLehn, 1990), geometry (Anderson, 1982), and physics (Larkin, 1981; VanLehn, Jones, & Chi, 1972). There also are several examples
of instructional systems that have been developed using the results and
ideas of cognitive models in computer-based tutoring systems (e.g., Anderson, Boyle, Corbett, & Seurs, 1990; Suppes & Morningstar, 1972) and
as resources providing teachers with information about student thinking
and learning (Carpenter, Fennema, Peterson, Chiang, & Loef, 1989).

Models of cognitive structures and processes in problem solving and
in learning have created connections between the study of processes and
the study of contents in education, particularly in mathematics and science. Figure 3 depicts my impression of the situation as it is now, with

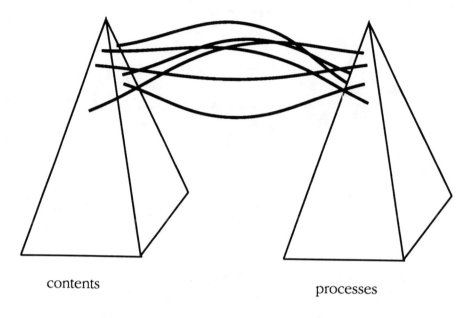

contents   processes

FIGURE 3.   Information-processing models have created some connections between studies of educational contents and learning processes.

the two bodies of study on contents and practices still largely separate, but with some threads linking them in various places. (Think of each thread as corresponding to a model of the processes of solving problems or of learning in some instructional domain, such as a model of solving geometry-proof exercises.)

The ability to develop information-processing models of routine instructional tasks is an underused scientific resource for educational improvement. If the society were to support a significantly expanded program of cognitive research and development, we could, in a period of about 10 years and at a cost of a few hundred million dollars, produce a collection of information-processing models for the standard routine problem-solving and question-answering tasks in the kindergarten through 12th-grade (K–12) curriculum. These models could be used in a number of ways, including having them as part of teacher education and incorporation of them into intelligent tutoring systems. A relatively complete set of cognitive models for routine tasks in the K–12 curriculum would be an important resource for education that could support significant improvements in the teaching of skills for performing those tasks.

It should be clear that constructing models of performance of routine tasks would contribute only indirectly to the important goals of educational reform that involve students becoming more active participants in the social processes of learning. Indeed, such models, if incorporated in currently dominant educational practices, could retard our progress toward significant educational reform by creating more efficient versions of passive learning. However, use of information-processing models would facilitate practices in which students would be more engaged in meaningful practices of learning communities. Models of routine skills could be used to develop systems in which those skills could be learned routinely, perhaps in a way similar to the use of practice rooms in music education. The learning and assessment of routine skills that now dominate classroom practices could become peripheral activities in which students could engage independently or in small groups, thereby allowing class time to be used in meaningful discussions of concepts and in collaborative work on significant problems and projects.

The opportunity to develop a relatively complete set of information-processing models for routine instructional tasks would enable researchers to integrate more strongly studies of contents and processes in education. This potential situation is illustrated by Figure 4, which shows strong integration across the content–process gap at the upper part of the two bodies of knowledge and practice. Connecting the upper parts of current knowledge about subject-matter contents and cognitive processes of school learning is a worthwhile goal.

Some Prospects

Researchers should also consider what would be involved in joining understanding of subject-matter and of cognition more fundamentally. The parts of the pyramids that can be connected can be viewed as outcomes of social practices of two kinds of intellective activities. On the contents side, the social practices are the activities of communities in which knowledge is generated in the society. On the learning-processes side, the social practices are the activities of communities of learners, mainly in schools.

JAMES G. GREENO

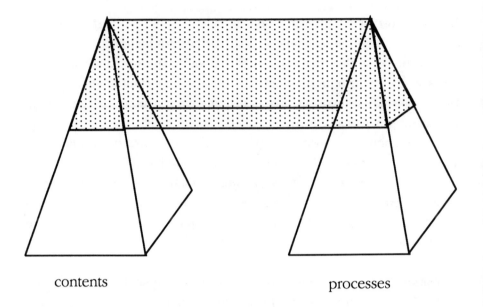

contents   processes

FIGURE 4. Information-processing models for routine tasks could be developed for the K–12 curriculum, providing an integrating bridge for studies of educational contents and learning processes.

To integrate the two bodies of study and research, I propose a focus on generative principles in those two social practices, especially on forms of discourse that are productive of knowledge and understanding. To represent this intuition in Figure 5, I have added a labeled dimension to Figure 4, with "routine question answering and problem solving" at the upper end and "practices of sense making and conceptual growth" at the lower end. Figure 5 expresses a claim that unified scientific understanding of contents and processes of mathematics and science education would be extended by a continuation of the study and reform of intellective practices of conceptual understanding and of conceptual growth, in communities that specialize in the theoretical inquiry of scholarly disciplines and in those that specialize in strengthening children's conceptual understanding and their abilities to use concepts and principles of subject-matter disciplines as intellective resources in their lives.

Components of the scientific effort to deepen synthesis of contents and processes in educational research are underway in the studies of

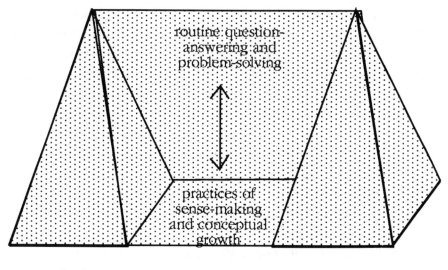

contents   processes

FIGURE 5. A deeper integration will be achieved with studies of practices of sense-making and conceptual growth.

communities of educational practices and of conceptual understanding and growth that I have surveyed briefly in previous sections of this chapter. Scientific understanding that connects more basic aspects of learning processes and contents will grow as researchers continue to investigate the character of activities in which people participate for making conceptual sense of events and systems in their lives, including their school lives, and ways in which conceptual understanding grows as children and adults become more adept in reasoning and discourse using various concepts.

Current research directions are not as well integrated as they need to become for the unification of processes and of contents to occur in a strong way. Well-developed methods and concepts currently used for studying processes of social interaction, especially in classrooms, do not provide much information about the contents of what people do and learn in those interactions. Studies of classroom discourse (Cazden, 1986), for example, provide very valuable characterizations of structures of participation in classroom interactions, addressing such questions as which members of a class make contributions, to whom the presentations of students and teachers are addressed, and which students' presentations are recognized and confirmed by teachers. Studies under the rubric of ethnomethodology (e.g., Garfinkle, 1967) and of conversation analysis (e.g., Sacks, 1974; Schegloff, 1991) provide detailed accounts of interactive processes in terms of structures of participation in activity. The methods and concepts available for analyzing patterns of social participation, however, provide little information about ways in which understanding of subject-matter or everyday concepts is involved in interactions or changes as a result of the interactions.

On the other hand, there are also well-developed methods and concepts for studying ways in which the contents of children's and adults' understanding and use of concepts change as they grow older or have more experience in domains of activity in which the concepts are involved. These studies of the growth of conceptual understanding and of cognitive expertise say little about processes that bring about more sophisticated understandings and uses of concepts.

Our understanding of processes and contents of learning needs to be unified, and there are different ways to do that. The approach that seems most promising involves attempting to unify our ability to study social interaction, especially discourse, with our ability to study the conceptual structures involved in reasoning and in problem solving. A current effort, in which I am engaged with Randi Engle and other students and associates, is to develop ways of analyzing conversational discourse that show how conceptual understanding is involved in the content of the discourse and how changes in conceptual understanding occur as an aspect of conversational processes. Our goal is to find ways to characterize conversations that include at least three sets of analytical features: (a) features that Clark and his associates (e.g., Clark & Schaefer, 1989; Clark & Wilkes-Gibbs, 1986) have developed to characterize conversations as structures of presentations and confirmations that construct informational common ground; (b) features that Sacks (e.g., 1974), Schegloff (e.g., 1991), and others have developed to characterize conversations as structures of interpersonal interaction in which participants coconstruct roles, commitments, and feelings, and (c) features developed in our studies of reasoning (Greeno et al., 1992) for characterizing conceptual understanding as attunement to constraints and to affordances that support inferences and representations in communities of practice. To the extent that we can succeed in bringing these three analytical strands together, we will have an analysis that allows us to understand conceptual contents of communicative and interpersonal processes in which conceptual understanding is achieved through constructing common ground among participants who are mutually committed to each other and to the understanding that they construct.

The view of research sketched in Figure 5 helps to clarify issues involving educational goals, as well as issues involving scientific directions. One of these is an apparent weakness of the position that favors educational reform that would make students more active participants in intellective practices rather than recipients of preformed knowledge. A strong intuition in this view is that students should be engaged in authentic practices of subject-matter disciplines to learn to use concepts and extend their understanding through activities of inquiry (e.g., J. S. Brown et al., 1989). Objections to this view often point out that the professional practices of subject-matter academicians seem unreasonable as models for nearly all students prior to graduate school (e.g., Palincsar, 1989; Wineburg, 1989).

I agree with the intuition that students' educational activities should be authentic practices of academic disciplines, as well as with the objection that professional practices of subject-matter academicians are inadequate as models for students. What is needed, it seems, is an understanding of the term *authentic practices* that captures the intuition but avoids the objection.

I propose, as a tentative framing of the problem, to focus on practices of conceptual inquiry, an idea that was central in Dewey's (e.g., 1938) discussions of educational goals. Inquiry is the main occupation of professional practitioners of academic disciplines, whose work is successful to the extent that progress is made in the understanding of important aspects of the material, mental, and cultural systems that constitute our lives and environments, including the development of more effective concepts and methods of communicating, of reasoning, and of formulating and solving problems. Inquiry should also be the main occupation of students as practitioners of the work of learning. Also, their work is successful to the extent that they progress in their understanding of important aspects of material, mental, and cultural systems, including becoming more effective in communicating, in reasoning, and in formulating and solving problems.

Within this fundamental similarity of the functions of professional academic work and of students' work, there also are fundamental differences. The practices of a community of professional academicians presuppose a body of knowledge, concepts, and methods in their discipline that is shared by the active participants and that the work of the community should extend. Reconstructing previously established knowledge, concepts, or methods may be necessary in some situations, but it is generally considered to result from failure of the community to attend sufficiently to results of previous work. The practices of a community of students presuppose that a body of knowledge, concepts, and methods is available as a resource that the members of this community have not mastered, and that their work succeeds to the extent that they become adept at knowing, understanding, and using the information, concepts, and methods of the discipline. If students' work adds significantly to the body of knowledge, concepts, and methods of a discipline, that is generally viewed as a wonderful event, but is not expected or required for students' work to be viewed as successful.

This suggests that development of an appropriate analysis of inquiry practices in different intellectual communities would resolve the dilemma of authentic educational activities. Practices that are appropriate, when knowing and understanding by the participants of the discipline's main results, concepts, and methods are assumed, should differ significantly from practices that are appropriate when the main activity is to increase the participants' knowing and understanding of results of previous disciplinary work. As an example, the role of teaching seems quite different in the two kinds of communities. Professional researchers teach each other, as well as prospective researchers, serving as mentors for activities that create new knowledge. A teacher of students in school is a mentor mainly for activities that use the information, concepts, and methods of a discipline as a resource by students for understanding, reasoning, and solving problems that occur in the present and future lives of participants of society.

Conclusion

Educational practice has been influenced profoundly by ideas developed in psychological research. Current practices in American education are shaped significantly by the ideas and methods of behavior analysis, with the organization of curricula and of testing influenced strongly by a view of knowledge as an accumulation of simple units combined in more complex structures (e.g., Gagné, 1965; Skinner, 1938; Thorndike, 1932). Science educators have been influenced by Piaget's ideas of cognitive development in which conceptual learning depends on the growth of general reasoning schemata (e.g., classification and conservation) and have therefore included activities designed to facilitate acquisition of those general structures. Soviet educators were much influenced by ideas of knowledge as sociocultural activity (e.g., Davydov, 1990; Vygotsky, 1962).

These three approaches in the psychology of learning are reflected in the three research topics discussed in this chapter. The study of cognitive structures is continuous with behaviorism in assuming that knowing is an organized accumulation of elementary components of procedures and declarative knowledge. The study of conceptual understanding and of conceptual growth is continuous with Piagetian studies in framing the study of children's school learning in larger patterns of cognitive development. The study of cognition and learning in communities of practices is continuous with sociocultural ideas of activity theory in emphasizing the social origin and character of concepts and of activities of learning. To the extent that current and prospective research efforts extend these various approaches and bring them together to form more integrated approaches, the development of long-standing intellectual traditions in psychology and their interrelations is continued.

I find it particularly appealing, however, to consider these current developments as continuous with the tradition of American pragmatism and functionalism in psychology and in philosophy. Advocating learning environments in which students participate in intellective activities that emphasize inquiry seems particularly continuous of Dewey's (1938) thinking; and emphasizing that learning occurs through participation in the social interactions of constructing meaningful concepts continues ideas developed by Mead (1934). Perhaps the strongest historical influence on current psychological thinking about education is a continuation of pragmatism, an influence that has been dormant in academic psychology for a few decades but that American psychologists carry because of our participation in the general culture of our society.

References

Anderson, J. R. (1982). Acquisition of cognitive skill. *Psychological Review, 89*, 396–406.

Anderson, J. R., Boyle, C. E., Corbett, A. T., & Lewis, M. T. (1990). Cognitive modeling and intelligent tutoring. *Artificial Intelligence, 42*, 7–49.

Barwise, J. (1989). *The situation in logic.* Stanford, CA: Center for the Study of Language and Information.

Barwise, J., & Perry, J. (1983). *Situations and attitudes.* Cambridge, MA: MIT Press.

Belenky, M. F., Clinchy, B. M., Goldberger, N. R., & Tarule, J. M. (1986). *Women's ways of knowing.* New York: Basic Books.

Briars, D. J., & Larkin, J. H. (1984). An integrated model of skill in solving elementary word problems. *Cognition and Instruction, 1*, 245–296.

Brown, A. (Chair). (1992, April). *Learning and thinking in a community of learners.* Poster symposium presented at the meeting of the American Educational Research Association, San Francisco, CA.

Brown, J. S. (1991, January–February). Research that reinvents the corporation. *Harvard Business Review*, 102–111.

Brown, J. S., & Burton, R. R. (1978). Diagnostic models for procedural bugs in basic mathematical skills. *Cognitive Science, 2*, 155–192.

Brown, J. S., Collins, A., & Duguid, P. (1989, January). Situated cognition and the culture of learning. *Educational Researcher, 18*, 32–42.

Burton, R. R. (1982). Diagnosing bugs in a simple procedural skill. In D. Sleeman & J. S. Brown (Eds.), *Intelligent tutoring systems* (pp. 157–184). San Diego, CA: Academic Press.

Cain, C. (1991). Personal stories: Identity acquisition and self-understanding in Alcoholics Anonymous. *Ethos, 19*, 210–254.

California State Board of Education. (1992). *Mathematics framework for California public schools.* Sacramento: California Department of Education.

Carey, S. (1985). *Conceptual change in childhood.* Cambridge, MA: MIT Press/Bradford Books.

Carpenter, T. P., Fennema, E., Peterson, P. L., Chiang, C.-P., & Loef, M. (1989). Using knowledge of children's mathematics thinking in classroom teaching: An experimental study. *American Educational Research Journal, 26*, 499–531.

Carraher, T., Carraher, D., & Schliemann, A. (1985). Mathematics in the streets and in schools. *British Journal of Developmental Psychology, 3,* 21–29.

Cazden, C. B. (1986). Classroom discourse. In M. C. Wittrock (Ed.), *Handbook of research on teaching* (pp. 432–463). New York: Macmillan.

Chi, M. T. H., Feltovich, P., & Glaser, R. (1981). Categorization and representation of physics problems by experts and novices. *Cognitive Science, 5,* 121–152.

Clark, H. H., & Schaefer, E. F. (1989). Contributing to discourse. *Cognitive Science, 13,* 259–294.

Clark, H. H., & Wilkes-Gibbs, D. (1986). Referring as a collaborative process. *Cognition, 22,* 1–39.

Cobb, P., Wood, T., & Yackel, E. (1990). Classrooms as learning environments for teachers and researchers. *Journal for Research in Mathematics Education, 4,* pp. 125–146.

Cognition and Technology Group at Vanderbilt. (1990). Anchored instruction and its relation to situated cognition. *Educational Researcher, 19,* 2–10.

Davydov, V. V. (1990). *Types of generalization in instruction: Logical and psychological problems in the structuring of school curricula* (Vol. 2). Reston, VA: National Council of Teachers of Mathematics.

DeKleer, J., & Brown, J. S. (1984). A qualitative physics based on confluences. *Artificial Intelligence, 24,* 7–83.

Devlin, K. (1991). *Logic and information.* Cambridge, England: Cambridge University Press.

Dewey, J. (1938). *Logic.* New York: Harper.

diSessa, A. A. (in press). Toward an epistemology of physics. *Cognition and Instruction Monographs.*

Dweck, C. S., & Legett, E. L. (1988). A social–cognitive approach to motivation and personality. *Psychological Review, 95,* 256–273.

Eckert, P. (1989). *Jocks and burnouts.* New York: Teachers College Press.

Eckert, P. (1990). Adolescent social categories—information and science learning. In M. Gardner, J. G. Greeno, F. Reif, A. H. Schoenfeld, A. diSessa, & E. Stage (Eds.), *Toward a scientific practice of science education* (pp. 203–216). Hillsdale, NJ: Erlbaum.

Eckert, P., & Knudsen, J. (1992). *Interdisciplinary research in support of science and mathematics education.* Palo Alto, CA: Institute for Research on Learning.

Fawcett, H. (1938). *The nature of proof.* New York: Teachers' College, Columbia University.

Gagné, R. M (1965). *The conditions of learning.* New York: Holt, Rinehart & Winston.

Garfinkle, H. (1967). *Studies in ethnomethodology.* Englewood Cliffs, NJ: Prentice Hall.

Gelman, R., & Gallistel, C. R. (1978). *The child's understanding of number.* Cambridge, MA: Harvard University Press.

Gibson, J. J. (1986). *The ecological approach to visual perception.* Hillsdale, NJ: Erlbaum. (Original work published 1979)

Greeno, J. G. (1978). A study of problem solving. In R. Glaser (Ed.), *Advances in instructional psychology* (Vol. 1, pp. 13–75). Hillsdale, NJ: Erlbaum.

Greeno, J. G., & Moore, J. L. (1993). Situativity and symbols: Response to Vera and Simon. *Cognitive Science, 17*, 49–60.

Greeno, J. G., Moore, J. L., & Mather, R. (1992). *Conceptual competence in situated reasoning about quantitative functional relations.* Stanford, CA: School of Education, Stanford University.

Hatano, G., & Inagaki, K. (1987). Everyday and school biology: How do they interact? *Quarterly Newsletter of the Laboratory of Comparative Human Cognition, 9*, 120–128.

Hutchins, E. (1990). The technology of team navigation. In J. Galegher, R. E. Kraut, & C. Egido (Eds.), *Intellectual teamwork* (pp. 191–220). Hillsdale, NJ: Erlbaum.

Hutchins, E. (1991a). The social organization of distributed cognition. In L. B. Resnick, J. M. Levine, & S. D. Teasley (Eds.), *Perspectives on socially shared cognition* (pp. 283–307). Washington, DC: American Psychological Association.

Hutchins, E. (1991b). *How a cockpit remembers its speeds.* Department of Cognitive Science, University of California, San Diego.

Hutchins, E. (in press). Learning to navigate. In S. Chaiklin & J. Lave (Eds.), *Understanding practice.* Cambridge, England: Cambridge University Press.

Inhelder, B., & Piaget, J. (1964). *The early growth of logic in the child* (E. A. Lunzer & D. Papert, Trans.). New York: Harper & Row. (Original work published 1959)

Johnson-Laird, P. N. (1983). *Mental models: Toward a cognitive science of language, inference, and consciousness.* Cambridge, MA: Harvard University Press.

Jordan, B. (1989). Cosmopolitical obstetrics: Some insights from the training of traditional midwives. *Social Science and Medicine, 28*, 925–944.

Kintsch, W., & Greeno, J. G. (1985). Understanding and solving word arithmetic problems. *Psychological Review, 92*, 109–129.

Klahr, D. (Ed.). (1976). *Cognition and instruction.* Hillsdale, NJ: Erlbaum.

Lampert, M. (1990). When the problem is not the question and the solution is not the answer: Mathematical knowing and teaching. *American Educational Research Journal, 27*, 29–64.

Larkin, J. H. (1981). Enriching formal knowledge: A model for learning to solve problems in physics. In J. R. Anderson (Ed.), *Cognitive skills and their acquisition* (pp. 311–335). Hillsdale, NJ: Erlbaum.

Larkin, J. H., McDermott, J., Simon, D. P., & Simon, H. A. (1979). Models of competence in solving physics problems. *Cognitive Science, 4*, 317–345.

Latour, B., & Woolgar, S. (1979). *Laboratory life: The social construction of scientific facts.* Beverly Hills, CA: Sage.

Lave, J. (1988). *Cognition in practice: Mind, mathematics, and culture in everyday life.* Cambridge, England: Cambridge University Press.

Lave, J. (1993). *Tailored learning: Apprenticeship and everyday practice among craftsmen in West Africa.* Manuscript in preparation.

Lave, J., Murtaugh, M., & de la Rocha, O. (1984). The dialectic of arithmetic in grocery shopping. In B. Rogoff & J. Lave (Eds.), *Everyday cognition: Its development in social context* (pp. 95–116). Cambridge, MA: Harvard University Press.

Lave, J., Smith, S., & Butler, M. (1988). Problem solving as everyday practice. In R. I. Charles & E. A. Silver (Eds.), *The teaching and assessing of problem solving* (pp. 61–81). Reston, VA: National Council of Teachers of Mathematics.

Lave, J., & Wenger, E. (1991). *Situated learning: Legitimate peripheral participation.* Cambridge, England: Cambridge University Press.

Lynch, M., & Woolgar, S. (Eds.). (1990). *Representation in scientific practice.* Cambridge, MA: MIT Press.

Marshall, H. (1972). Structural constraints on learning. In B. Geer (Ed.), *Learning to work.* Beverly Hills, CA: Sage.

Martin, L. M. W., & Scribner, S. (1988). *An introduction to CNC Systems: Background for learning and training research.* New York: Graduate School and University Center of the City University of New York.

Mathematical Sciences Education Board. (1990). *Reshaping school mathematics: A philosophy and framework for curriculum.* Washington, DC: National Research Council, National Academy of Sciences.

McCabe, V., & Balzano, G. J. (Eds.). (1986). *Event cognition: An ecological perspective.* Hillsdale, NJ: Erlbaum.

Mead, G. H. (1934). *Mind, self, and society.* Chicago: University of Chicago Press.

National Council of Teachers of Mathematics. (1989). *Curriculum and evaluation standards for school mathematics.* Reston, VA: National Council of Teachers of Mathematics.

Palincsar, A. S. (1989, May). Less charted waters. *Educational Researcher, 18,* 5–7.

Piaget, J., Grize, J., Szeminska, A., & Bang, V. (1977). *The psychology and epistemology of functions* (F. X. Castellanos & V. D. Anderson, Trans.). Dordrecht, The Netherlands: D. Reidel. (Original work published 1968)

Resnick, L. B., Bill, V., & Lesgold, S. (in press). Developing thinking abilities in arithmetic class. In A. Demetriou, M. Shayer, & A. Efklides (Eds.), *The modern theories of cognitive development go to school.* London: Routledge & Kegan Paul.

Riley, M. S., & Greeno, J. G. (1988). Developmental analysis of understanding language about quantities and of solving problems. *Cognition and Instruction, 5,* 49–101.

Sacks, H. (1974). An analysis of the course of a joke's telling in conversation. In R. Bauman & J. Sherzer (Eds.), *Explorations in the ethnography of speaking* (pp. 337–353). Cambridge, England: Cambridge University Press.

Saxe, G. (1990). *Culture and cognitive development: Studies in mathematical understanding.* Hillsdale, NJ: Erlbaum.

Scardamalia, M., & Bereiter, C. (1991). Higher levels of agency for children in knowledge building: A challenge for the design of new knowledge media. *Journal of the Learning Sciences, 1,* 37–68.

Schegloff, E. A. (1991). Conversation analysis and socially shared cognition. In L. B. Resnick, J. M. Levine, & S. D. Teasley (Eds.), *Perspectives on socially shared cognition* (pp. 150–171). Washington, DC: American Psychological Association.

Scribner, S. (1984). Studying working intelligence. In B. Rogoff & J. Lave (Eds.), *Everyday cognition: Its development in social context* (pp. 9–40). Cambridge, MA: Harvard University Press.

Simon, D. P., & Simon, H. A. (1978). Individual differences in solving physics problems. In R. Siegler (Ed.), *Children's thinking: What develops?* Hillsdale, NJ: Erlbaum.

Skinner, B. F. (1938). *Behavior of organisms.* New York: Appleton-Century-Crofts.

Suchman, L. (1987). *Plans and situated actions.* Cambridge, England: Cambridge University Press.

Suppes, P., & Morningstar, M. (1972). *Computer-assisted instruction at Stanford, 1966–1968: Data, models, and evaluation of the arithmetic programs.* San Diego, CA: Academic Press.

Thorndike, E. L. (1932). *Fundamentals of learning.* New York: Teachers College, Columbia University.

Turvey, M. T. (1992). Ecological foundations of cognition: Invariants of perception and action. In H. L. Pick, Jr., P. W. van den Broeck, & D. C. Knill, (Eds.), *Cognition: Conceptual and methodological issues* (pp. 85–117). Washington, DC: American Psychological Association.

VanLehn, K. (1990). *Mind bugs: The origins of procedural misconceptions.* Cambridge, MA: MIT Press/Bradford Books.

VanLehn, K., Jones, R. M., & Chi, M. T. H. (1972). A model of the self-explanation effect. *Journal of the Learning Sciences, 2,* 1–60.

Vygotsky, L. S. (1962). *Thought and language.* Cambridge, MA: MIT Press.

Wellman, H. M. (1990). *The child's theory of mind.* Cambridge, MA: MIT Press/Bradford Books.

Wenger, E. (1990). *Toward a theory of cultural transparency: Elements of a social discourse of the visible and the invisible.* Palo Alto, CA: Institute for Research on Learning.

Wineburg, S. (1989, May). Remembrance of theories past. *Educational Researcher, 18,* 7–10.

Yates, J., Bessman, M., Dunne, M., Jertson, D., Sly, K., & Wendelboe, B. (1988). Are conceptions of motion based on a naive theory or on prototypes? *Cognition, 29,* 251–275.

PART THREE

Social and Cultural Factors in Mathematics and Science Education

Social and Cultural Factors in Mathematics and Science Education

George M. Batsche

Those who school America's children and youth face challenges unmet in the history of American education. The public schools in the United States now have the responsibility for educating a larger number of students from a greater diversity of backgrounds and with a wider range of abilities than at any point in history. The basis of the nation's economy has moved from one of manufacturing and production to that of information management and service, and the need for good education is of the greatest importance in such an economy. However, according to the professional literature and the public press, the probability of getting a "good education" is declining rapidly, particularly for minority students and for students from low-income families. The general concern and dissatisfaction with the schooling of America's children and youth was well articulated in *A Nation at Risk* (National Commission on Excellence in Education, 1983), *A Place Called School* (Goodlad, 1984), and *High School* (Boyer, 1983). Since these publications appeared, there has been a national reaction and response to the allegations that America's schools were not preparing students to understand the new world order, to compete successfully in a global economy, and to profit from living in a world of increasing cultural diversity. The most visible product of this national concern came from the initial work of the National Education Goals Panel and the Bush administration in the form of the national education plan, *America 2000* (U.S. Department of Education, 1991).

The six education goals (and objectives) contained in *America 2000* (see Gallagher, chapter 1 this volume) attempt to address concerns regarding the education crisis in America. As Penner notes in the Introduction to this volume, however, the goals address desired outcomes but do not provide the mechanisms for achieving those outcomes. In particular, mechanisms are not provided to address issues related to increasing graduation rates of minority students and to increasing the numbers of women and minority students who succeed in mathematics and science. What is most troubling about this omission is the implication that the mechanisms either are not known or, if known, that they could be applied equally to all students. Moreover, to achieve the goals contained in *America 2000*, schools must recognize that raising standards alone will only further serve to alienate students at risk for educational failure. What is needed, and not provided, is a basic understanding of the mechanisms necessary to engage at-risk students longer and more successfully in the process of schooling. This is particularly true in the areas of mathematics and science education, both for women and for students from minority and low-income backgrounds. The fact that the mechanisms to reduce this at-risk status are not in place is reinforced by data from the 1992 Metropolitan Life Survey of the American Teacher. The results of that survey indicate that approximately 55% of all teachers (elementary, junior high, and high school) feel that students are unprepared to learn at their grade level and that more than three fourths of teachers who work in schools that serve predominantly minority and low-income students feel that the students are not prepared to learn at their grade level. The percentages remain consistent across all grade levels. If mechanisms were in place to address successfully these obstacles to learning, then one would expect to see the percentage of teachers who say that their students are unprepared to learn decline as grade level rises. This is not the case.

Clearly, schools are not addressing these obstacles for students in general. However, those most affected are minority students and those from low-income families.

In his recent book, *The Predictable Failure of Educational Reform*, Seymour Sarason (1990) identified a number of societal and systemic issues that contribute to his belief that we are unable to explain the historical "intractability" of schools to respond to reform efforts. He cited the underestimated effects of "daunting heterogeneity," curricula devoid of interest and meaning, ineffective power relationships, and the fact that the in-school world suffers (by comparison with the out-of-school world) in terms of interest, stimulation, and motivation as possible sources of this intractability. Sarason (1990) asserted that to truly begin the process of reform and to respond to the heterogeneity in the student population while facilitating productive power relationships,

> you start with their worlds. You do not look at them, certainly not initially, as organisms to be molded and regulated. You look at them to determine how what they are, seek to know, and have experienced can be used as the fuel to fire the process for enlargement of interests, knowledge, and skills. You do not look at them from the perspective of a curriculum, class-room, or school structure. You enter their world to comprehend and rein-force the psychological assets they already possess. (p. 164)

Perhaps the members of the American Psychological Association's (APA's) President's Task Force on Psychology in Education (Spielberger, 1992) had Sarason's perspective in mind when they developed and published a draft of *Learner-Centered Psychological Principles* (see McCombs, chapter 9 this volume). The reader who reviews these principles will, at once, see Sarason's concepts nested within their framework. At this writing, there are 12 principles encompassing topics that include metacognitive and cognitive, social, developmental, and individual-differences factors. These principles are based on research by psychologists and educators in the areas of learning, motivation, and human development. Some of the most exciting work in learner-centered principles is being conducted in areas that relate directly to the *America 2000* goals of increasing graduation rates of minority students and of improving achievement in mathematics and science. The learner-centered principles were derived from the work of individuals such as the authors of the chapters in this section. Therefore, the reader will note a great deal of similarity between the concepts expressed by the learner-centered principles and those brought out in the chapters. For that reason, it is worth presenting some of the learner-centered principles discussed at length in chapter 9:

> *Principle 1*: Learning is a natural process that is active, volitional, and internally mediated: It is a goal-directed process of constructing meaning from information and experience, filtered through each individual's unique perceptions, thoughts, and feelings.

> *Principle 5*: The depth and breadth of information processed, and what and how much is learned and remembered, is influenced by (a) self-awareness and beliefs about one's learning ability (personal control, competence, and ability); (b) clarity and saliency of personal goals; (c) personal expectations for success or failure; (d) affect, emotion, and general states of mind; and (e) the resulting motivation to learn.

When schools are organized around a particular set of principles based on a narrow set of values (White, male, middle class) that are not reflective of the learner, then schools are likely to fail certain kinds of students. For example, it has been argued that when the process of schooling is conducted in competitively based learning environments, then many students (often female students and those from non-White, non-middle-class backgrounds) do not thrive and are at risk for poor performance and disidentification. Students who feel unwelcome, uncomfortable, or incompetent either leave the school setting willingly (drop out), unwillingly (are suspended or expelled), or remain physically but are not committed or involved. In this context, Sykes (1984) introduced the concept of "the Deal." The Deal may be made with a few students, a class, or an entire school. In essence, the Deal refers to an unwritten, but understood, agreement that a student will receive a passing grade if he or she does not cause trouble, does not question too much, and is relatively compliant. The Deal could also include the teachers not expecting too much from female students (compared with male), Black students (compared with

Whites), or poor students (compared with rich). Students feel that they are not respected in this environment and that the teacher or system has already abandoned them. There is a significant loss of capacity and productivity under such circumstances, yet such circumstances are common in our educational system. The Deal may form the basis for initiation of the process of disidentification discussed by Jones in this section. The concept of the Deal may be applied to teachers and to school administrators as well. Teachers survive the years before tenure by recognizing that the use of traditional teaching methods, establishing "control" in the classroom (possibly at the expense of individual student learning), teaching the curriculum instead of the learner, and engaging in behavior that supports the system help to ensure their longevity. It is anticipated that minority students and those from low-income families will not do as well in school settings and that female students will not do as well in mathematics and science. If this happens, then the prophecy is fulfilled and the Deal is made. The Deal survives when the focus is on the concerns of the system, not on the learner. Unless the *America 2000* goals incorporate mechanisms such as those proposed by the learner-centered principles, more "deals" will be struck with those at risk for educational failure, and the process of schooling will result in success for fewer and fewer students.

The authors of the chapters in this section suggest specific mechanisms for dealing with cultural differences in education (Jones); gender differences in mathematics ability, anxiety, and attitudes (Hyde); and gender differences in cognitive style (McGuinness).

James Jones reviews the research on cultural influences on learning and reinforces many of the concepts set forth in the learner-centered principles. He emphasizes the point that learning and academic performance are associated with the extent to which the learning context takes into account the psychosocial context of the learner. Jones addresses the underachievement of African–American children and explores the research seeking to explain why these children underperform in comparison with White children in academic areas. Through an explanation of disidentification, the recognition of home–school discontinuities, the effects of internalizing inferiority, and the impact of Black English vernacular,

Jones seeks to provide an understanding that the underachievement of African–American students is not related to child-centered "deficiencies" but to a range of possible explanations for which mechanisms of change exist. These explanations, along with the effects of ethnic socialization, differential expectations for academic success, and cultural mistrust, form the basis of psychosocial approaches to educational change. Jones reviews a number of models and interventions that use an ecological social perception approach to learning and presents suggested principles of education reform that show promise for the improved performance of Black children. This chapter provides a perspective on cultural influences on learning that offers to the reader a greater understanding of the issues that African–American students face in the process of schooling.

In chapter 7, Janet Hyde presents the issues related to gender differences in mathematics ability, anxiety, and attitudes from a meta-analytic research perspective. She discounts the concept of large, general differences in the performance of men and women in mathematics. Hyde argues for a small effect size overall, with differences noted as a function of level of schooling and type of mathematics skill. She presents the dilemma (maximalist vs. minimalist) of how the data are to be interpreted. Should the data supporting no gender difference (in computation, understanding mathematics concepts at any age, or problem solving in the early grades) using a minimalist approach or the data supporting a moderate difference (in problem solving beginning at high school) using the maximalist approach be used when citing implications for mathematics education? The most interesting points in this chapter may lie in Hyde's use of her data to develop implications for education. She explores how these "believed" large gender differences are communicated to students and how communication affects academic performance.

Guidance counselors, parents, and others who influence the schooling of students and take the maximalist approach will guide female students away from the mathematics and science areas. Hyde integrates the effects of the maximalist approach with course taking, with course content review, and with when the test is taken to provide a possible explanation for lower SAT scores for female students. She also explores how teaching styles and classroom organization can result in more "girl-friendly" classrooms. She applies the maximalist–minimalist argument to gender differences in mathematics anxiety as well as to mathematics ability. In the chapter, Hyde does not limit her discussion of gender differences to students in public elementary and secondary schools. She extends the effects of adopting the maximalist approach to the university setting. Hyde explores issues (in addition to the ones at the elementary and secondary levels) such as the lack of role models and overt discrimination (including sexual harassment) to explain why fewer than 10% of the university faculty in mathematics and sciences are women.

Hyde concludes by providing a framework around which to build one's own perspective on gender issues in mathematics and science and provides a rationale for looking beyond the issue of gender difference to provide possible answers for the schools of the future.

In chapter 8, Diane McGuinness takes issue with Hyde's conclusion that few differences exist between men and women in mathematics ability. She looks beyond the more global outcome differences between men and women and examines in detail the different strategies that each gender uses to solve mathematical problems. McGuinness feels that gender differences research has been "overly politicized." She therefore begins the chapter by presenting and discounting a number of sociocultural theories to explain the differences, including the social conspiracy model and the cultural accident model. As an alternative hypothesis to the environmental theories, McGuinness suggests that males and females adopt different strategies or styles in learning mathematics concepts. She presents the research on gender differences in spatial thinking and interest (objects vs. persons) as a model for the types of differences seen in mathematics between the genders. McGuinness integrates the research on gender differences in spatial ability and interest with the research on different types of learning styles in a discussion of the development of mathematics strategies. She presents the strategies used by male and female students through the developmental period of the schooling process and identifies how the different strategies used by the genders results in differential performance as a function of the mathematics skill required. This discussion picks up where Hyde's leaves off. Although McGuinness's approach might be characterized as maximalist in nature, she is really attempting to provide an explanation for the *specific* differences that one sees between the sexes, regardless of the magnitude of that difference. Her work and that of the individuals whom she cites in the chapter provide a fertile ground from which to draw conclusions about restructuring classrooms and curricula in mathematics and science. Drawing on the work of Piaget and Davidson, McGuinness presents a number of important questions to be answered in seeking solutions to the gender differences in mathematics and science performance. McGuinness concludes by offering a challenge to create classrooms that use a radically different approach to teaching mathematics skills—one that might eliminate gender differences in mathematics. Given the way in which schools are structured, and the power of the Deal, those who choose to accept this challenge have a formidable task. However, McGuinness provides them with a great deal to think about.

Each of the authors in this section reviews relevant research, offers alternative conceptualizations, and issues challenges to the reader to explore alternatives freely and to "think outside of the box." Each author provides support, indirectly, for the use of the learner-centered principles in creating mechanisms to ensure equity in educational outcomes for students at risk for all types of educational failure. Significantly, each of these authors writes about inclusionary strategies designed to maximize the effects of schooling for all students. The future economic growth of this country lies in its ability to enable all of its citizens to perform at their best. The vehicle to achieve that performance is the schooling process. The application of sound psychological principles to education, as delineated in the following chapters, is the best solution to create a schooling process that results in equal outcomes for all who participate, regardless of gender or ethnic affiliation, and eventually to eliminate the Deal.

References

Boyer, E. L. (1983). *High school: A report on secondary education in America.* New York: Harper & Row.

Goodlad, J. A. (1984). *A place called school.* New York: McGraw-Hill.

National Commission on Excellence in Education. (1983). *A nation at risk: The imperative for educational reform.* Washington, DC: Author.

Sarason, S. B. (1990). *The predictable failure of educational reform.* San Francisco: Jossey-Bass.

Spielberger, C. D. (1992). APA is developing learner-centered psychological principles. *Communique, 21,* 15–18.

Sykes, G. (1984). The deal. *Wilson Quarterly, 7,* 59–77.

U.S. Department of Education. (1991). *America 2000: An education strategy.* Washington, DC: Author.

CHAPTER 6

Psychosocial Aspects of Cultural Influences on Learning Mathematics and Science

James M. Jones

To paraphrase the English poet John Donne, no child is an island; no child stands alone. Children's thoughts, strivings, needs, and choices emerge from a social context at once vivid and strong, and confusing and daunting. In a school, a child is not a passive receptacle who memorizes important facts on command, and strives to please the teacher as a means of enhancing self-esteem or, at least, of assuaging superego pressures. A child is an active organism guided in part by his or her own attempts to chart a viable course through the murky waters of maturation and development on the one hand, and of multiply influential social contexts that simultaneously open and close options and opportunities on the other.

In the late 1960s, I was living in Connecticut. One weekend, a friend who taught in a Harlem elementary school in New York City brought three of her students out for a getaway to the country. They were excited and filled with energy and wonder. One of the young boys, Reggie, and I engaged in a foot race around the house. We started at the same point but ran in opposite directions. The winner was the one who returned to "go" first. I was well ahead as we passed on the opposite side of the house, but as we passed, Reggie fell. I stopped and went over to see if he was alright. He jumped up as I approached and resumed the race to a successful conclusion. I was impressed by his guile and pragmatism. That evening over dinner, after the children had gone to bed, I learned that when our friend asked them what they would want if they could have one wish granted, Reggie had said that he would like a new brain so that he could be smart!

We have led children to believe that their academic performance is simply a matter of brainpower. Poor performance can be remediated, in Reggie's mind, only with a brain transplant. The fact is that performance is multiply determined and that contextual effects, accumulated over time, broadly define the critical role of culture in schooling outcomes. It is the role of cultural influences that is the focus of this chapter.

The defining premise of this chapter is that how children learn mathematics and science is linked closely to the broad psychosocial world in which they develop and adapt. Within this perspective, schooling is the institutionally organized medium for learning that stands as the locus or confluence of multiple aspects of individual development, community resources and socialization, traditions and philosophies of education, and specific curriculum content and pedagogical style. Learning is a complex process that implicates all of these factors.

The material reviewed in the following pages will show how much difference in learning can be accounted for by psychosocial factors not generally identified in the learning process. Strategies for improving learning outcomes generally focus on the interactive elements among the learner, the teacher, the content, and the home–family–community elements. The idea that there are facts and/or skills that teachers teach to students who can learn them is a model that is fast giving way to the more complex psychosocial dynamics represented in the approaches described in this chapter.

Ceci (1991) argued that IQ levels are related directly to the amount of time that a child spends in school. If school is dissociated from the psychosocial dynamics that children must negotiate as they develop, it seems unlikely that spending more time in these situations would improve their performance. However, if the psychosocial goals are met, then learning may improve as well as performance. In this situation, it is significant to realize that the more time children spend in appropriate learning situations, the better their educational outcomes will be. The current, well-replicated finding that African–American children decline in performance relative to Whites as they move through the educational system suggests that they spend less time in school, relative to Whites, and that the time spent there is less conducive to learning.

In their early work on learning and thinking among the Kpelle of Liberia, Cole, Gay, Glick, and Sharp (1971) observed a fundamental relationship, "People will be good at doing the things that are important to them and that they have occasion to do often"(p. xi). They demonstrated this by showing that when they attempted to measure memory capacity among the Kpelle, they found them to be less proficient than their American counterparts, even when specific attempts were made to assist them by providing structural guides that had proven effective in studies in America. Rather than concluding that they had deficient capacity for memory, Cole et al. carried the experiment further. They embedded memory objects in a folk narrative that was meaningful for the Kpelle subjects. The result of this methodological modification was that free-recall memory was profoundly affected both in its extent and in its organization. Tracing the items dropped by a chieftain's daughter as she is being spirited away by a bad bogeyman resulted in higher recall rates and significant object order correlations that mirrored the story line. When memory tasks were linked to meaningful social context and to required modes of thinking that were more familiar, performance improved significantly.

The implication of this research and of this general line of reasoning is clear: Learning and academic performance will be associated with the extent to which the learning context takes into account the psychosocial context of the learner. To the extent to which this is not done, it may be no surprise that learning suffers and performance declines. This performance decrement is at the heart of the problem that this chapter seeks to address.

What Is the Problem?

The core problem that this chapter addresses is the underachievement of African–American children. No matter what age, economic circumstance, or subject matter is involved, African–American children underperform relative to White children. The National Assessment of Educational Progress (National Center for Educational Statistics, 1985) estimates the gap to be 10%–18% in reading, mathematics, and science.

Entwisle and Alexander (1990) showed that, at the beginning of first grade, Black and White children do not differ in verbal or mathematics computation skills. However, Black children are behind White children in reasoning skills. Stephenson, Chen, and Uttal (1990) also documented mathematics underperformance in Black and Hispanic children in the first, third, and fifth grades. However, when mother's education levels were controlled, these differences disappeared in the fifth grade. In spite of the poorer performance of minority children, minority mothers and their teachers valued education and a variety of specific measures to improve educational outcomes (e.g., more homework, competency based testing, longer school days) more than their White counterparts.

These performance gaps are all the more significant because of the population trends projected for the coming years. Ethnic minorities will constitute more than 30% of the United States population by the year 2010 (U.S. Bureau of the Census, 1990). They will likely constitute an even higher percentage of the school-age population. The prediction that, by the year 1995, only 15% of the new people entering the workforce will be White males makes the successful and comprehensive education of ethnic minorities and women a matter of practical necessity (Johnston & Packer, 1987).

Why do these performance gaps persist? Is it biological determinism? Psychological maladaptation? Is it cultural incompatibility? Racism in the schools? Is it some combination of all of these? In the following section, I consider a variety of explanations for the performance gap that are distinguished by their emphasis on psychosocial and cultural factors. In the following section, I discuss briefly approaches to learning that offer positive models for using cultural components effectively in learning strategies. In a similar vein, the next section considers approaches to educational change that show promise for improving the educational performance of African–American children. This discussion concludes with a review of several principles by which learning approaches to education changes may be informed by psychological and sociocultural models.

Explanations for Academic Underachievement of African–American Children

Most approaches to explaining the lower academic performance of minority children acknowledge psychosocial or cultural factors, or both. They vary in orientation primarily by the relative emphases that they place on family socialization, school-linked practices, philosophies and structures, characteristics of the learner, the curriculum and the pedagogical style, peer relations, community adaptations, and cultural styles that have evolved over years of adaptation.

This section provides an overview of the range of explanations that derive primarily from the perspective of African–American researchers. It should be kept in mind that these approaches are not mutually exclusive, that is, it appears that each identifies an aspect of the learning context that would improve if that perspective were adopted.

Disidentification

One approach was proposed by Steele (1988) on the basis of his self-affirmation theory. According to self-affirmation theory, when an important self-concept is threatened, an individual's primary self-defensive goal is to affirm the integrity of the self, not to resolve the particular threat. As a result, adaptation to a specific self-threat is subsumed by more general motivations to affirm either a broader self-concept or a different, but equally important, aspect of the self-concept.

Steele applied this reasoning to the case of Black academic achievement to explain persistent underachievement (Steele, 1992b). He argued that when faced with persistent and expected devaluation in the academic arena, the resultant expected self-concept vulnerability leads Black students to *disidentify* with the academic domain of self, substituting either more global self-referents or equally important alternate domains (e.g., sports, social activism). By disidentifying with academic performance domains, one can avoid the self-concept implications of underperformance by bolstering self in other arenas. These self-affirming alternatives allow one to maintain positive identity in the face of expected vulnerabilities in an important self domain. Disidentification insulates one from the ego-threat of race bias in academic life.

One of the important implications of this analysis is that it is the adaptive response to stigma and prejudice that is implicated in underperformance, not ability. Some evidence for this is suggested by the observation that if Black and White students are matched on SAT scores, Black students will still have lower cumulative grade point averages than comparably scoring White students. Steele (personal communication, April 20, 1992) also looked at these effects experimentally. College undergraduates were presented with one of two tests, described either as diagnostic of their academic ability, or extremely challenging and difficult. All students performed better when the test was described as challenging than when it was presented as diagnostic of their abilities; and there were no differences in performance between Black and White students when the test was described in this manner. However, when the test was described as diagnostic of ability, Blacks performed at a lower level than Whites. Do Black students believe that they have less ability, or that a test that purports to measure it will be inherently biased against them? Whatever the explanation, reactions to expected and persistent stigmatization may well be a significant contributing factor to academic performance of Black students.

School–Home Discontinuities

Schooling is a socializing process that has historically served as an avenue to upward mobility for those who achieve success. It has also been systematically denied to or undermined for African–American children. Boykin and his colleagues (Allen & Boykin, 1991; Boykin, 1983; Boykin & Allen, 1988) have taken the view that underachievement of African–American children could be understood to be a result of classroom approaches that fail to appreciate the cultural context from which children come. Specifically, Boykin argued that there is a "cultural deep structure," which informs schooling concepts and processes, that is at variance with the cultural capital brought by African–American children to the school environment. The result of these home–school discontinuities is a pattern of schooling that perpetuates underperformance and looks to the child or to his or her home, or community environment to place blame. Boykin argued that, when the child's cultural integrity is both acknowledged and taken into account in the structure of learning, performance improves dramatically. This basic orientation leads to four characterizations of school settings that, if acknowledged and modified in certain ways, could ameliorate the performance decrements of African–American children. These are

1. A culture of power dominates conventional approaches to schooling that work to the detriment of Black children, thus alienating them from learning itself and from the communities of practitioners.
2. Schools play a socialization role that is discrepant from the real or desired opportunities available to Black children.
3. Schools have failed to appreciate the integrity of the Black cultural experience and how it might influence or be used by the schooling process.
4. Schools have failed to develop continuities with the out-of-school experiences of African–American children.

These elements conspire to undermine the quality of education for African–American children. Research by Boykin et al. showed that performance improves for African–American children, but is not affected for White children, when rhythmic-movement opportunities accompany learning tasks (Allen & Boykin, 1991; Boykin & Allen, 1988), tasks are presented in variable and stimulating ways (Boykin, 1982; Tuck & Boykin, 1989), and learning situations are structured around a communal or collective rather than individualistic and competitive orientation (Ellison & Boykin, in press).

Internalizing Inferiority Undermines Academic Performance

Howard and Hammond (1985) suggested that the performance gap can be explained, in part, by the following sequence of cognitions and behaviors: (a) internalizing societal prescriptions of intellectual inferiority, (b) inferring an internal or genetic basis for this intellectual inferiority, and (c) tending to avoid intellectual competition. This analysis tracks that of Steele's (1988) disidentification model.

Howard and Hammond (1985) offered an attributional analysis in which internalized inferiority leads to a performance-expectancy processes by which one comes to expect behavioral failure that reinforces cognitive representations. People attribute failure to internal causes. Thus, expectancy leads to a self-fulfilling prophecy, which, in turn, confirms the expectancy and reinforces the attributions of low ability.

They argued for (a) controlling expectancy communications so as to establish more positive expectancies, (b) developing an "intellectual work ethic" in the Black community so as to make intellectual processes a valued element in the community of practitioners, and (c) altering attributional styles so as to explain successes by internal causes (ability) and failures by external ones (effort).

Black English Vernacular Effects on Mathematics Education

Eleanor Orr is the founder of the Hawthorne School in Washington, DC. Founded in 1956, the high school drew heavily from middle- and upper-middle-class White families. In 1972, the school was invited by the DC Board of Education to enter into a cooperative arrangement that would transfer 41 students, chosen by lot, from the public schools. Over a 9-year period, 320 students enrolled in Hawthorne, of whom 98% were Black. In the first 2 years, 87% of the Black students failed the mathematics and science courses in which they had enrolled.

In 1987, Orr published a book titled *Twice as Less*, in which she made the intriguing observation that mathematics and science failures of Black students, relative to White students, could be traced to their Black English vernacular (BEV) usage. Whereas it had long been argued that BEV is a language with real linguistic properties and that it deserves to be taught in schools as a means of improving the reading potential of Black students, Orr theorized that certain language features may influence mathematics and science performance.

The basic postulate is that the use of prepositions in BEV is such that certain quantitative relationships cannot be expressed in ways that accurately represent them in standard English vocabulary. As a result, Black students at Orr's school had difficulties reaching correct answers because they used language rules and practices that inhibited their successful performance. This performance was not related to diligence or intellectual ability, but to socially constructed linguistic patterns.

Consider the following example. In standard English, the word *than* expresses comparisons that are additive or subtractive (John traveled two more miles *than* Sam; Sam traveled two fewer miles *than* John). By contrast, the *as* mode can be used for multiplicative or partitative comparisons (John traveled twice as many miles *as* Sam; or Sam traveled half as many miles *as* John). The grammar provides what is true mathematically. Moreover, when the multiplier is greater than one, the comparison may be expressed in either mode (as in the previous example), suggesting that either addition or multiplication can be used to reach the answer. However, when the multiplier is a fraction, the *than* mode cannot be used because the quotient of a division cannot be reached by subtraction. Orr was able to show through case examples that her BEV-speaking students often combined these modes, thereby reaching linguistic conclusions that did not mirror the mathematical facts.

Thus, language is not only a source of cultural connectivity and of social intercourse, but in relationship to schooling and societal standards for performance, it may also be a critical dimension defining access and accomplishment in mathematics and science. This further complicates the picture because, whereas language may be an avenue to facilitate intrapersonal dynamics leading to general improvement in school-based reading (as was decided in Ann Arbor in 1979; cf. Smitherman, 1991), this assertion of Orr's, if true, broadly implicates major features of culture in learning and successful performance in mathematics and science.

Ethnic Socialization Effects

Moore (1986) evaluated family socialization effects directly. She administered the Wechsler Intelligence Scale for Children (WISC) to 46 socially defined Black children, half of whom were adopted by Black families and half by White families. All had been adopted by 2 years of age. Results showed significantly higher WISC scores for the transracially adopted children than for the traditionally adopted ones (117 vs. 104).

In a second phase, conducted in the children's homes, Moore (1986) observed strategies used by the mothers to help their children perform well on a cognitively complex block design task from the Wechsler Belle-vue Scale. Moore found that White mothers tended to release tension through joking and laughing, give more positive evaluations of their child's performance (e.g., "You're good at this"), give more generalized hints to help in problem solving, and show more enthusiasm in the form of cheers and applause. Black mothers, by contrast, tended to release tension in more negative ways such as scowling or frowning, give more negative evaluations of their child's performance (e.g., "You know that doesn't look right"), give specific instructions rather than general hints for prob-lem solving, and generally use expressions of displeasure at the child's performance (e.g., "You could do better than this if you really tried").

The significant differences in IQ in the absence of any biological relationship between children and adoptive parents makes the patterns of interaction of great interest. The pattern of observed differences is provocative and in need of follow-up. It seems to suggest that beliefs about child rearing, teaching styles, and affective expression differ be-tween Black and White parents. The question one must ask is, Why? Because these two sets of adoptive parents were similar in socioeconomic status, although not identical because they differed both in paternal ed-ucation levels and in overall socioeconomic index, there is a temptation to look to ethnic/racial variations for the answers. It appears that the Black mothers were less generous in their praise, affective support, and encouragement. This kind of observation demands more systematic re-search. Follow-up interviews might reveal the foundation of their attitudes about parenting, teaching, and socialization that gives rise to these be-havioral differences. There may well be real and persistent differences in attitudes, norms, and beliefs that translate to behavioral differences. If these differences have systematic effects on academic performance, this would be important to know.

To carry these environmental effects even further, Plomin and Daniels (1987) evaluated the question of why children in the same family are so different. They argued that three factors account for differences among children: biology, nonshared environmental effects, and shared environmental effects. The first relates to how much variance can be accounted for by genes, the second to how much variance is accounted for by the fact that the children are influenced by different environmental events, and the third to the effects of shared environments. They concluded that nonshared environmental effects are responsible for differences in personality, temperament, and IQ (after childhood). This means that, even though parents may think that they are imparting similar values, attitudes, and standards to their children and providing them with generally similar experiences, what will affect their children most is what they experience uniquely, not what they share with their siblings.

Differential Expectations for Academic Success

Ross and Jackson (1991) were interested in the persistent pattern of lower levels of academic achievement by Black male students, relative to Black female students. They noted that between 1976 and 1981, the number of Black female students earning bachelor's degrees increased by 9%, whereas the number of Black male students decreased by 9%. During the same time frame, the number of Black women earning the PhD degree increased by 29%, whereas the number of Black men earning the PhD decreased by 10%. Ross and Jackson speculated that the spiraling problems of African–American males may well create a self-fulfilling prophecy that is fueled by teachers' lowered expectations for the academic performance of these young men.

To assess this idea, they asked 90 suburban New York schoolteachers (kindergarten through sixth grade) to complete a questionnaire containing 12 identical case histories that varied on academic achievement (good, fair, poor), submissiveness (submissive, nonsubmissive), and sex (male, female). Twenty-nine teachers rated the scenarios for year-end success, future success, and desirability to have in class. Results showed that teachers (a) tended to like girls better; (b) did not distinguish on submissiveness with regard to achievement, but did prefer submissive children in class; (c) liked the nonsubmissive Black boys the least; and (d) predicted least likelihood of future success for nonsubmissive Black boys.

Cultural Ecology of Competence

Ogbu (1985) proposed a *cultural–ecological* model of competence among inner-city Blacks. This model is offered as an alternative to both the traditional *universalist* model, which assumes that successful school performance is due to abilities possessed by the child which result, in part, from middle-class parental socialization; and the *difference* model, which suggests that poor performance of Black children results from the failure of schools to appreciate the integrity of Black cultural instrumental competencies. The *cultural–ecological* model proposes that Black children's competencies derive from the adaptation to the environmental demands and adult expectations for their behaviors.

Specifically, the cultural–ecological model assumes that (a) "child-rearing is a culturally organized formulae [*sic*] to ensure that newborns survive to become competent adults who will contribute to the survival and welfare of their social group" (Ogbu, 1985, pp. 49–50); and (b) "the environment which influences childbearing and development is much broader than is usually defined in current developmental studies . . . [and] . . . must be expanded to include the nature of cultural tasks faced by the population, and the way in which these tasks determine the competencies that are transmitted and acquired" (p. 51).

When applied to inner-city Black Americans, the cultural–ecological model is heavily influenced by the idea that one must adapt to an economy that can be characterized by "a scarcity of jobs; dead-end, peripheral, and unstable jobs; and by low wages and little social credit as measured by the values of the larger society" (Ogbu, 1985, p. 19). Moreover, these circumstances are not new, but reflect years of discrimination and repression to which ecological adaptation has been tuned. The cultural competencies, then, may be thought of as survival strategies designed to create opportunity and possibility from circumstances that provide relatively little of either. What is implied by this analysis is that reasonable adaptations to the ecological contingencies of inner-city Black children may lead to the development of behaviors, skills, and preferences that are at variance with those assumed by the pedagogical approaches to learning. To then blame children, families, and communities for the "failure" to achieve academically is, by this reasoning, a double insult.

Cultural Mistrust

Mistrust is one of the legacies of race relations in this society. Terrell and his associates (F. Terrell, Terrell, & Taylor, 1981; T. Terrell & Terrell, 1984; Watkins & Terrell, 1988; Watkins, Terrell, Miller, & Terrell, 1989) developed the Cultural Mistrust Inventory (CMI), designed to measure the degree to which Blacks mistrust Whites in a variety of situations. Some sample items are "It is best for Blacks to be on their guard when among Whites," "Whites will say one thing and do another," and "A Black person can usually trust his or her White co-workers." Scores on this inventory have been related to several potentially important variables. One interesting study proposed that cultural mistrust might have the effect of lowering motivation to cooperate or to perform in situations managed or controlled by Whites. The authors (F. Terrell et al., 1981) reasoned that IQ test scores could be influenced by the extent to which one were willing to perform for a White experimenter when level of mistrust was high. One hundred Black college students who had scored high or low on the CMI were administered the Wechsler Adult Intelligence Scale (WAIS) by a Black or White graduate student.

It was hypothesized and found that subjects who scored high on the CMI and who were tested by a White examiner did the poorest on the WAIS. Specifically, WAIS scores for high-CMI students were 95.6 when the examiner was Black, but only 91.7 when the examiner was White. Conversely, for low-CMI students, WAIS scores were only 86.4 for Black examiners, and 97.8 for White examiners. Not only was there a lowering of performance for White relative to Black examiners in the high-mistrust circumstance, there was an elevation for White and a depression for Black in the low-mistrust condition. This latter observation raises the interesting possibility that when students are not sensitized to the political and personal implications of racial discrimination and bias, they are subject to a form of in-group derogation and out-group favoritism that has corresponding consequences for their performance. Whether this interpretation is true or not, it is important to note the influence of cultural perceptions on performance when these cultural attitudes and values are stimulated by interracial contexts.

Conceptualizing Psychosocial Concepts in Learning

The approaches summarized in the preceding section share the view that learning processes and outcomes cannot be dissociated from the cultural contexts in which learning takes place. The attitudes, values, beliefs, adaptations, social perceptions, and symbolic representations all enter the dynamic linkages among teachers, learners, and material or processes to be learned. Although this viewpoint often takes on political connotations when social or educational policy is addressed, it is increasingly being understood as a core feature of the learning paradigm itself.

In this section, I review briefly two models of learning that integrate these viewpoints in their conceptions of the learning process itself. The following two models, Greeno's (1989) *situated learning* and Lave's (1991) *community of practice*, argue that the processes and the performance outcomes must meet context demands and characteristics if successful learning is to take place. These approaches provide another perspective on the recurring theme of this chapter.

Situated Learning

Greeno (1989) offered a perspective on thinking in which he noted that scientific progress in understanding performance on specific tasks has exceeded progress in understanding the psychology of critical, productive, higher order, and creative thinking. The relative lethargy of progress in these theories of thinking can be traced, according to Greeno, to three fundamental framing assumptions:

1. *Situated cognition*, in which the locus of thinking is assumed to be in the mind, rather than in interaction between an agent and a physical or social situation.

2. *Personal and social epistemologies,* in which processes of thinking and learning are assumed to be consistent across persons and situations. Beliefs and understandings about cognition differ across people and social groups, and properties of thinking are determined by these contexts.

3. *Conceptual competence,* in which resources for thinking are assumed to be knowledge and skills built from simple components, rather than general capabilities resulting from everyday experience and native endowment. Learning and thinking are activities in which children elaborate and reorganize knowledge rather than simply applying and acquiring cognitive structures and procedures.

Cognition is a relation between an actor and a situation, in which the actor brings a set of beliefs and understandings about learning that derive from his or her personal experience and social constructions, and in which performance is expressed as the elaboration and reorganization of this knowledge and understanding. Intelligence is not a commodity that can be claimed by simply outputting specific knowledge in response to an input trigger. Rather, it is the output of a process that may depend on a host of factors connected with the active learner.

This approach emphasizes experiential and interactive learning. It also places importance on the beliefs that a learner brings to the situation: One must believe that one can learn in order to do so. Teachers must also recognize that learners bring tacit conceptual competencies to the learning situation. There is ample evidence that both aspects of this education are not met. That is, teaching does not always take the learner's perspective into account (e.g., teacher expectancy) and the learner often does not feel that his or her own resources are appropriate or relevant to the learning task.

Conceptualizing learning or thinking in this way demands that certain aspects of culture be represented not simply as a politics of curriculum reform, but as an essential element of theory, research, and pedagogy of learning. The "cultural capital" (Delpit, 1988) that a child brings to the learning process is a deficiency that "prevents" him or her from learning, but potentially is an asset that will aid his or her learning if the learning process can effectively take it into account.

Knowledgeable Skills in a Community of Practice

Lave (1991) introduced the notion of communities of practice in the following way:

> Socially shared cognition does not simply result in the acquisition or internalization of knowledge, but in becoming a member of a sustained community of practice. Developing an identity as a member of a community and becoming knowledgeably skillful are part of the same process, with the former motivating, shaping, and giving meaning to the latter. (p. 65)

In American society, race continues to be a salient and significant feature of social significance. Thus, to a great degree, one's racial group, perhaps delimited by socioeconomic circumstances, becomes the *community of practice*. To the extent that this is true, then understandings and beliefs about cognition may show a divergence between the school setting and the community setting. Lave also talked about participation in the community and moving from newcomer to old-timer status. Peripheral learning occurs when a newcomer is a part of the community, but at a peripheral level. The newcomer learns through secondary methods, and teaching may not, in fact, be perceived to be taking place. To the extent that schooling is not a legitimate part of the community of practitioners, then the situation is not one in which learning can take place, at least the kind of motivated learning implied by the situating of learning in a community of practice.

If the community of practice is antithetical to the formalized learning setting, either one accepts, in the abstract, that the formal setting is an alternate route to the community of practice, or one acknowledges a desire to move to a different community of practice. This may be what Steele (1988) called the domain of performance.

Lave distinguished *knowledgeable skill* from knowledge. Knowledge may be commoditized, objectified, and seen as an object of the learning process. Institutions are in business to transmit these commodities. Knowledgeable skills are defined by the community of practitioners. Discontinuities between the institutions and the communities marginalize the learner, and lead to the formation of interstitial communities within the institutions. These people may, in fact, have gained more knowledgeable skills, if not mastered the knowledge products of the institutions.

Psychosocial Approaches to Educational Change

The preceding two sections have considered a variety of explanations for academic underperformance of African–American children, and models by which some of these reasons can be conceptually incorporated into the emerging learning paradigms. The dynamics of learning, thinking, and perception are heavily influenced by social context.

Just as Greeno and Lave built these dynamics into their models of learning, McArthur and Baron (1983; see also Zebrowitz, 1990) offered an explicit model for the ecological context of social perception. Four principles describe their ecological theory. First, social perception serves an *adaptive function* either for the survival of the species or goal attainment of individuals. Second, perception is directed by the properties of the environment called *affordances* (environmental properties of things guide perception but do not define it). Third, the environment is structured so that its affordances (the way in which things should be viewed) are revealed: Over time, events come to define what objects afford. A first encounter may not reveal whether a person can be trusted or not (is trustworthy), but over time, the structure of events comes to tell us whether the property of trustworthiness is an accurate one. Fourth, the detection of social affordances is determined in part by *attunements* of the perceiver, that is, the stimulus information to which the perceiver attends.

If this ecological theory is taken as a model of how individuals form knowledge and think about events in their environment, and formulate perceptual and behavioral strategies for dealing with them, one can see that this ecological approach has merit as a means of describing the social context effects in learning. The following models and interventions can be understood to employ certain principles of the ecological social perception approach to the dynamics of learning.

Getting Smart: Efficacy Training

Howard (1990) adapted the situated learning conceptions to a program of efficacy training that he called "Getting Smart." This view rejects the premise that intelligence is innate and rejects the educational processes that seem to follow from this view, such as ability tracking. The negative effects of this innate viewpoint include feelings of inferiority and lowered self-esteem, expectation for academic success, and feelings of personal control. Getting Smart seeks to reverse these tendencies by adopting an alternative approach based on the notion that "development [is] a process of building capacity . . . [and that] . . . all children can learn, if the process of learning is effectively organized and managed by adults" (p. 11).

Getting Smart is based on three critical elements:

1. Teaching children a constructive theory of development, such as "smart is not something that you just are; smart is something that you can *get*." Getting smart is a student's choice and, in this model, should be taught early on at home and in the school. The foundation of this approach is nurturing intellectual development as an ongoing process of building analytic and operational capability through effort.

2. Building children's confidence through belief and emotional support. This approach emphasizes positive communications that generate positive attitudes in the child about his or her capacity to learn. It rests primarily on the affirmation of the teacher or authority with responsibility for the child's learning outcomes. The defining characteristic of this process is the child's belief that "I am the kind of person who can learn whatever is taught to me in school" (p. 13).

3. Teaching children the efficacy of effective effort, step by step, that is, beginning by choosing a starting point that matches the difficulty of the material with the present capabilities of the student. The learning goal should be challenging, but attainable; both failure and success should be possible outcomes. As learning progresses, goals become more challenging as capabilities increase. Over time, the material is more complex and the child is "smarter."

The efficacy program has been tested in the Detroit public schools where the principle of "all students can learn" is put into practice with the step-by-step paradigm outlined earlier. Preliminary results suggest that children in the efficacy program, in the third and sixth grades, performed better on both the verbal and mathematics portions of the California Achievement Test than did a control group who did not participate in the program (Howard, 1990).

Schooling, Culture, and Home Environment

Boykin (in press) suggested that schools are deeply embedded in the cultural fabric of society. They have evolved from an early 20th-century ethos that does not apply now and will clearly not apply in the 21st century. The superficialities of cultural diversity fail to capture the deep cultural significance of schooling in America, which needs to be overhauled. He proposes five specific changes:

1. Schools should focus on pedagogy more than on socialization to a cultural standard.
2. Schools should be more client centered rather than fulfilling organizational mandates that are culturally driven.
3. Schools should emphasize talent development (cognitive growth and learning potential) rather than talent assessment.
4. Schools should acknowledge the diversity of society in a genuine multicultural education program.
5. Education should be functional for the realities of the 21st century.

Proposed remedies include reconceptualizing schooling along the lines suggested earlier. The critical aspect of Boykin's claims is the integrity of African–American culture and its implications for teaching pedagogy. He argued that it is possible to delineate specific aspects of Afrocultural expressions that have meaning for learning contexts. He found one element, movement, to have implications for learning.

Boykin and Allen (1988) had 80 low-income Black children, aged 5 to 9 years, learn picture pairs in one of two conditions: high movement, during which a percussive tune was played and children clapped hands during the acquisition; and low movement, during which simple rote-memorization techniques were used with children sitting in their seats. Children learned the pairs better in the high-movement condition than the low-movement one.

In a follow-up study (Allen & Boykin, 1991), the acquisition phase was similar, but the testing phase was conducted with music present or absent. Subjects were Afro- and Euro-American children aged 4 to 8 years. Low movement had subjects sit still and practice a recitation procedure for learning the picture pairs. In high movement, children stood around the experimenter and listened to music while learning the pairs. During testing, music was present or absent. Results showed that White children did better under low-movement acquisition conditions but that Black children did better under high-movement conditions. However, these differences were obtained only under no-music test conditions.

These findings suggest that stimulation may play a greater role in the acquisition of knowledge for African–American children than for White children. They also suggest that the gains in acquisition may not carry over to the performance output. Boykin (1986) argued that higher stimulation levels in the home may well require higher stimulation levels in the classroom. Rather than trying to suppress activity in the classroom, this reasoning goes, one should try to channel it into productive learning paradigms.

School Development Program

The school intervention program that has come to be known as "the Comer method" was begun in 1968 as a program of school improvement and preventive psychiatry by the Yale University Child Study Center in New Haven, Connecticut, under the direction of psychiatrist James Comer (Comer, 1980, 1989). Comer and his colleagues created a school–community subsystem of two elementary schools that taught 99% Black children, 70% of whom were from low-income, single-parent families. The interventionists lived in the community to learn firsthand what issues affected the schooling of these children. They learned quickly that

> the social context of learning was being inadequately addressed in schools, which in turn created problems that made teaching quite difficult, regardless of the methods. The content quickly became meaningless to many students.... Most important, we realized that educators paid little to no attention to child development issues, and such issues directly affected the

context of schooling. It was obvious that our intervention had to address the social context of schools. (Comer, 1989, p. 269)

The program that emerged from this intervention is called the School Development Program (SDP). It consists of three component programs:

1. *School Planning and Management Team (SPMT)*. This team is headed by the principal and consists of teachers (selected by peers), parents (selected by parents group), and a mental health person. The SPMT coordinates all the activities and resources of the school.

2. *Mental Health Team (MHT)*. This team consists of a social worker, a psychologist, and a special education teacher, and is headed by a member of the team or the principal. It helps to solve problems of individuals, but more important, it plays a prevention role by spotting potential or sources of problems and by trying to change the situation.

3. *Parents Program (PP)*. This team supports the social calendar of the school, participates in the governance, and helps teachers in classrooms, and generally members serve as role models, which helps children to form positive attachments and bonds with their parents or with other adults from the community.

In addition to New Haven, evidence from Benton Harbor, Michigan, and Prince George's County, Maryland, show that test performance is up (California Achievement Tests), suspensions are down, and self-esteem is growing. Again, social context provides a backdrop for schooling interventions and classroom modifications with positive results.

Wise Schooling

Steele (1982) proposed that the antidote to the negative academic consequences of the disidentification discussed earlier is to incorporate principles of *wise schooling* in the academic domain. Schooling is unwise when

1. It requires assimilation to a White mainstream that includes mastering the culture and ways of the American mainstream and giving up any particulars of a Black life-style.
2. It demands assimilation but fails to recognize or acknowledge the contributions of the groups that are expected to assimilate.

3. It stigmatizes members of these groups by assuming intellectual inferiority that will require remediation.

Against this set of requirements, expectations, and beliefs, a Black student may well disidentify with the educational enterprise.

Steele (1982) suggested that a wise approach to schooling would reverse these tendencies by adopting an educational goal based on four fundamental approaches:

1. Students must feel valued by the teacher for their potential as persons.
2. The challenge and the promise of personal fulfillment, not remediation (under any guise), should guide education.
3. Racial integration is a useful, but not a necessary, element of the wise schooling design.
4. The particulars of Black life and culture (e.g., art, literature, politics, and music) must be presented in the mainstream curriculum of American schooling, not consigned to special days, weeks, or months, or to special topic courses and programs aimed at Blacks.

This proscription for schooling Black students is offered as a means of reducing the psychological vulnerabilities that ethnic minorities face in the school environment, and of releasing the students to the empowering idea that they can and will succeed. Steele (1992a) reported on preliminary findings from a wise schooling program that targeted talented African–American college students. Initial analyses of the impact of this program on the students' academic achievement indicated that wise schooling may be effective in improving minority students' grades and bringing their overall academic performance into line with their general abilities.

Summary

The preceding discussions lead to some common ground. Children must feel that they are appreciated, valued, and loved by persons in authority and with responsibility for their education. They must feel that they can accomplish the educational goals set out for them or that they set for themselves. Most important, they must feel that their teachers believe that they can accomplish these goals and will work with them in developing their abilities to reach the goals. The learning process is not defined by the acquisition of facts, but by the creation of skills that enable children to take control of their own learning capacities, rates, and outcomes. The learner cannot be disentangled from his or her community because (a) the skills that have currency are generally defined by that community; (b) the significant socialization patterns that establish educational expectancies and therefore influence educational outcomes come from there; and (c) specific psychosocial patterns of behavior that interact with the classroom's pedagogical style emerge from the community from which children come.

Some Principles of Education Change

The preceding discussion has reviewed several perspectives on the problems of underperformance of Black students and has outlined the strategies and assumptions of several intervention projects. Undoubtedly, there will be many more intervention projects mounted to stem the tide of educational underperformance and failure in minority children. This chapter suggests strongly that these problems must not be viewed as residing solely with the children but as a matter of mismatches, misconceptions, and miscommunications among all parties to the learning process.

In closing this chapter, I discuss briefly some of the overarching concepts or principles that might be extracted from the foregoing discussions. These principles are offered as additional conceptual, methodological, and empirical context for understanding the general significance of the psychosocial factors in learning.

Multiple Levels of Context Effects

Social context effects operate on several levels, each of which presents problems and possibilities for action. At the *individual* level, one must confront self-awareness, self-esteem, self-concept, beliefs, and schemata or cognitive structures that impose meaning on the learning or schooling process. These cognitive structures of self can play vital roles if they include the notions of "I am capable," "My community thinks it important for me to learn," "My teachers want me to learn," and so on. However, if there are negatively balanced cognitive structures such as "I am dumb," it is very difficult to undo them.

For example, Lepper, Ross, and Lau (1986) had high school students enrolled in basic mathematics courses solve deductive reasoning problems with the aid of 10-minute instructional films. One film was substantially better than the other, and students who viewed that film performed substantially better. After students had received feedback about their performance, they were told of the different quality of the films and that the films had determined their performance. In spite of this feedback, students continued to draw unwarranted inferences about their own ability that were in line with their initial performance. This effect persisted and was even heightened at the 3-week follow up assessment. This finding led the authors to conclude that

> overcoming the pernicious effects of early school failures on students' self-perceptions and attitudes may prove a difficult assignment ... demonstrating to a child, even in a clear and concrete fashion, that his or her poor performance may well have been the consequence of an inept or biased teacher, a substandard school, or even prior social, cultural, or economic disadvantages may have little impact on his or her feelings of personal competence or potential. (p. 490)

This demonstration seems to support the ideas offered by both Steele and Howard in their analyses.

A second level is the *institutional* level in which socialization processes, role models, and expectancy effects can all operate to set non-achieving goals. More specifically, specific classroom practices may disadvantage certain children. Boykin (in press) suggested, most strongly, that patterns of pedagogy may fail to take into account preferred learning styles. If children arrive in the classroom with certain forms of "cultural capital," then institutional practices that fail to take them into account will disadvantage the young learners. The School Development Program model of Comer (1989) tries specifically to modify the institutional arrangements by bringing parents into the school and by creating teams of teachers, principals, psychologists, and social workers to coordinate school activities that take multiple contexts into account.

It may useful here to distinguish between *learning* and *schooling*. Learning takes place in a schooling environment. It is possible to learn in a schooling environment that is suboptimal, but it is not easy. The social context arguments apply, however, to both learning and schooling effects. Learning requires a somewhat more microlevel analysis of individual cognitive structures and specific interactional patterns between a student and teacher. Schooling implies broader social context factors involving the community, educational policy, curriculum, administration, and social organization.

A third level is the *cultural* level of community beliefs, knowledgeable skills, and values. Children arrive at school with the imprint of a cultural socialization that may be at variance with the standard pedagogical styles and expectations of the teachers, principals, and others. Ogbu (1985) was explicit in suggesting that cultural competencies transmitted at home and in the community may confer different skills than traditional schooling assumes. Failing to assess carefully the cultural standards to which ecological adaptation responds will likely doom educators to continue to offer schooling that fails certain children. Boykin's (in press) notion of "cultural deep structure" captures this same idea in a slightly different way. He argued that there are unstated cultural assumptions that children can detect and react against. These assumptions of powerlessness, lack of ability, and negative expectations establish a psychological orientation in children that will diminish their interest in and felt capability at learning. As Boykin (in press), Steele (1992), and Howard (1990) all demanded, the messages from teachers and the schooling environment in general must be positive and supportive.

Variability Within and Between Racial/Ethnic Groups

Perhaps it is obvious, but one must acknowledge that context effects vary within ethnic and racial groups, as well as among them. Failing to consider gender differences among Black students would surely limit the success of any educational intervention. The attempt to take these differences into account in a specific way is shown by all-male schools in Milwaukee,

Baltimore, and elsewhere. Although many decry this trend as an infringement of rights and a return to segregationist philosophy (e.g., Freiburg, 1991), it is fair to say that it is exactly the kind of sensitivity that may well prove to hold the greatest chance of success. Similarly, region of the country, urban–rural–suburban distinctions, socioeconomic differences, and ethnicity including language variations all must be factored into the broadest conception of social context.

Importance of Culture

Another important point is that culture does indeed matter. This is obvious also, but we have yet to determine what the real cultural differences are and which of these differences matter in the educational process and how. Writers vary in the extent to which they prescribe different educational practices and standards for members of different cultural groups. We need to conduct cultural inventories of behaviors, values, beliefs, skills. If we believe that there is a cultural capital that children bring to the learning situation, and that this capital is linked to principles of ecological adaptation and survival as Ogbu suggested, then we must systematically evaluate it and introduce the results to our schooling philosophy and practice.

Children Learn!

It is definitely the case that children learn. They may not learn what educators, teachers, and parents want them to learn, but they do learn. If the structure of formal learning in school is not perceived as possible or desirable, then it is going to lead to low rates of learning of school material. It will likely be linked to higher rates of learning of alternative skills and knowledge. Thus, it is not useful to talk about children's failures to learn. It would seem much more appropriate to consider why they fail to learn what we want them to learn.

Schooling as a Political Socialization Process

School represents values that, for many children, are at variance with those that evolve in their daily lives. School has historically been an agent of negative identities for African–American children. The stigma associated with racism is embedded in the structure and culture of schooling. It will take more than busing children to predominantly White schools to change this cultural deep structure of racism. It will also take a great deal to reclaim the school from the alternative value system that views schooling as a low-probability avenue to success. For many, racial identity is compromised by schooling. Schooling success seems to imply "acting White" for many African–American students (Ogbu & Fordham, 1986). This notion is not an educational but a political idea. As such, it becomes a basis for the politicization of racial and personal identity. It obviously interacts with principles of schooling and educational practice. Just as the subtleties of racism carry political implications and consequences, so, too, the "color blind" philosophy carries the implication that one may do well if one "loses" his or her racial identification. This politicization of identity becomes linked to a bicultural frame of reference within which being American or mainstream is juxtaposed with being Black. The two orientations are pitted against one another, with educational instrumentalities associated with the mainstream orientation. Therefore, to achieve educationally seems to imply that one is rejecting one's African–American frame of reference. This duality dilemma (Jones, 1991) surfaces as an identity conflict.

Major and Crocker (1992) offered another sort of ambiguity that derives from this duality, the *attributional ambiguity*. They addressed situations in which a stigmatized, target Black person is evaluating the causes of behavior of a nonstigmatized person directed at himself or herself. The behavior of the nonstigmatized person may be in response to the actual individual characteristics, or behavior of the stigmatized person, or it may reflect the stereotypes and prejudices that the nonstigmatized person holds. When the behavior is negative, it may be discounted as racially motivated, thus obscuring actual performance deficiencies. If it is positive, it may again be discounted as an example of sympathy or guilt. In the positive case, it could also have augmentation effects so that the performance is judged even better because the positive feedback occurred in spite of the prejudices that may exist.

Crocker, Voelkl, Testa, and Major (1991) provided support for this reasoning in a study in which Black and White college students received feedback from a White same-sex evaluator as part of a friendship development exercise. Subjects sat behind a one-way mirror with the blinds up or down. Black subjects attributed both the positive and negative feedback that they received to qualities of themselves more when the blinds were up than when they were down. White subjects did not show this tendency. Specifically, when Black subjects received negative feedback and were not seen, their self-esteem ratings declined. However, when they were seen, there was no decrease in their self-esteem ratings.

Thus, if a child feels stigmatized by virtue of his or her racial group membership, the child may *disidentify* with the learning context, or *discount* negative performance feedback by attributing it to the stigmatizing biases. Both of these processes would likely reduce the effectiveness of classroom teaching and of student learning outcomes.

Conclusion

This review suggests that the underperformance of Black children in educational settings is linked to psychosocial and behavioral consequences of social contexts in which learning takes place. Affective reactions to stigma and negative expectancies, adaptations to ecological demands, racist cultural assumptions of schooling, and discrepancies between the home-community environments and schooling practices contribute substantially to the poor learning and performance outcomes for Black children.

Programs to remedy the situation, as well as conceptions of learning, thinking, and the effects of social contexts on both, represent promising directions for improved performance by Black children. Consideration of the multiple levels at which context effects influence learning, the consequences of between- and within-group variations, the importance of culture and the need to know more about its effects, and the sociopolitical processes that occur in schooling situations suggest some of the specific principles that educational changes should take into account. There is a growing awareness of the psychological processes that influence learning and what can be done to incorporate them in programs to provide the kind of education that children need and deserve, and that our country increasingly requires.

References

Allen, B. A., & Boykin, A. W. (1991). The influence of contextual factors on Afro-American and Euro-American children's performance: Effects of movement opportunity and music. *International Journal of Psychology, 26*, 373–387.

Boykin, A. W. (1982). Task variability and the performance of Black and White school-children: Vervistic explorations. *Journal of Black Studies, 12*, 469–485.

Boykin, A. W. (1983). The academic performance of Afro-American children. In J. Spence (Ed.), *Achievement and achievement motives* (pp. 321–371). San Francisco: Freeman.

Boykin, A. W. (1986). The triple quandary and the schooling of Afro-American children. In U. Neisser (Ed.), *The school achievement of minority children* (pp. 57–92). Hillsdale, NJ: Erlbaum.

Boykin, A. W. (in press). Harvesting culture and talent: African–American children and educational reform. In R. Rossi (Ed.), *Educational reform and at-risk students*. New York: Teachers College Press.

Boykin, A. W., & Allen, B. A. (1988). Rhythmic movement facilitation of learning in working class Afro-American children. *Journal of Genetic Psychology, 149*, 335–348.

Ceci, S. J. (1991). How much does schooling influence general intelligence and its cognitive components? A reassessment of the evidence. *Developmental Psychology, 27*, 703–722.

Cole, M., Gay, J., Glick, J. A., & Sharp, D. W. (1971). *The cultural context of learning and thinking: An exploration in experimental anthropology.* New York: Basic Books.

Comer, J. P. (1980). *School power: Implications of an intervention project.* New York: Free Press.

Comer, J. P. (1989). The school development program: A psychosocial model of school intervention. In S. L. Berry & J. Asamen (Eds.), *Black students: Psychosocial issues and academic achievement* (pp. 264–285). Newbury Park, CA: Sage.

Crocker, J., Voelkl, K., Testa, M., & Major, B. (1991). Social stigma: The affective consequences of attributional ambiguity. *Journal of Personality and Social Psychology, 60*, 218–228.

Delpit, L. D. (1988). The silenced dialogue: Power and pedagogy in educating other people's children. *Harvard Educational Review, 58*, 280–298.

Ellison, C., & Boykin, A. W. (in press). Comparing outcomes from differential cooperative and individualistic learning methods. *Social Behavior and Personality.*

Entwisle, D. R., & Alexander, K. L. (1990) Beginning school math competence: Minority and majority comparisons. *Child Development, 61*, 454–471.

Freiburg, P. (1991, May). Separate classes for Black males? *APA Monitor*, pp. 1, 47.

Greeno, J. S. (1989). A perspective on thinking. *American Psychologist, 44*, 134–141.

Howard, J. (1990). *Getting smart: The social construction of intelligence.* Lexington, MA: Efficacy Institute.

Howard, J., & Hammond, R. (1985). Rumors of inferiority: The hidden obstacles to Black success. *The New Republic, September 9*, 17–21.

Johnston, W. B., & Packer, A. H. (1987). *Workforce 2000: Work and workers for the 21st century*. Indianapolis, IN: Hudson Institute.

Jones, J. M. (1991). The politics of personality: Being Black in America. In R. Jones (Ed.), *Black psychology* (3rd ed.). Berkeley, CA: Cobb & Henry.

Lave, J. (1991). Situating learning in communities of practice. In L. B. Resnick, J. M. Levine, & S. D. Teasley (Eds.), *Perspectives an socially shared cognition* (pp. 63–84). Washington, DC: American Psychological Association.

Lepper, M. R., Ross, L., & Lau, R. R. (1986). Persistence of inaccurate beliefs about the self: Perseverance effects in the classroom. *Journal of Personality and Social Psychology, 50*, 482–491.

Major, B., & Crocker, J. (1992). Social stigma: The consequences of attributional ambiguity. In D. M. Mackie & D. L. Hamilton (Eds.), *Affect, cognition and stereotyping: Interactive processes in group perception*. San Diego, CA: Academic Press.

McArthur, L. Z., & Baron, R. (1983). Toward an ecological theory of social perception. *Psychological Review, 90*, 215–247.

Moore, E. (1986). Family socialization and the IQ test performance of traditionally and transracially adopted Black children. *Developmental Psychology, 22*, 317–326.

National Center for Educational Statistics. (1985). *The condition of education*. Washington, DC: U.S. Government Printing Office.

Ogbu, J. (1985). A cultural ecology of competence among inner city Blacks. In M. B. Spencer, S. K. Brookins, & W. R. Allen (Eds.), *Beginnings: The social and affective development of Black children* (pp. 45–66). Hillsdale, NJ: Erlbaum.

Ogbu, J., & Fordham, S. (1986). Black students' school success: Coping with the "burden of acting White." *Urban Review, 18*, 176–206.

Orr, E. W. (1987). *Twice as less: Black English and the performance of Black students in mathematics and science*. New York: Norton.

Plomin, R., & Daniels, D. (1987). Why are children in the same family so different from one another? *Behavioral and Brain Sciences, 10*, 1–60.

Ross, S. I., & Jackson, J. M. (1991). Teachers' expectations for Black males' and Black females' academic achievement. *Personality and Social Psychology Bulletin, 17*, 78–82.

Smitherman, G. (1991). Talkin and testifyin: Black English and the Black experience. In R. Jones (Ed.), *Black psychology* (3rd ed., pp. 249–268). Hampton, VA: Cobb & Henry.

Steel, C. (1988) The psychology of self-affirmation: Sustaining the integrity of the self. *Advances in Experimental Social Psychology, 21*, 261–346.

Steele, C. (1992a, October). Invited address, annual meeting of the Society of Experimental Social Psychology, San Antonio, TX.

Steele, C. (1992b). Minds wasted, minds saved: Crisis and hope in the schooling of Black Americans. *Atlantic Monthly, 269(4)*, 68–78.

Stephenson, H. W., Chen, C., & Uttal, D. H. (1990). Beliefs and achievement: A study of Black, White and Hispanic children. *Child Development, 61*, 508–523.

Terrell, F., Terrell, S., & Taylor, J. (1981). Effects of race of examiner and cultural mistrust an the WAIS performance of Black students. *Journal of Counseling Psychology, 49*, 750–751.

Terrell T., & Terrell, S. (1984). Race of counselor, client sex, cultural mistrust level, and premature termination from counseling among Black clients. *Journal of Counseling Psychology, 31*, 371–375.

Tuck, K., & Boykin, A. W. (1989). Task performance and receptiveness to variability in black and white low-income children. In A. Harrison (Ed.), *The 11th Conference on Empirical Research in Black Psychology*. Washington, DC: NIMH Publications.

U.S. Bureau of the Census. (1990). *Statistical abstracts of the United States: 1990* (110th ed.). Washington, DC: Author.

Watkins, C. E., & Terrell, F. (1988). Mistrust level and its effects on counseling expectations in Black client–White counselor relationship: An analogue study. *Journal of Counseling Psychology, 35*, 194–197.

Watkins, C. E., Terrell, F., Miller, F., & Terrell, S. (1989). Cultural mistrust and its effects on expectational variables in Black client–White counselor relationships. *Journal of Counseling Psychology, 36*, 447–450.

Zebrowitz, L. A. (1990). *Social perception*. Pacific Grove, CA: Brooks/Cole.

CHAPTER 7

Gender Differences in Mathematics Ability, Anxiety, and Attitudes: What Do Meta-Analyses Tell Us?

Janet Shibley Hyde

Hundreds of studies exist on gender differences in mathematics performance and ability and on gender differences in mathematics attitudes and affect. Moreover, several relevant meta-analyses exist summarizing the results of these studies. In the sections that follow, I review a major meta-analysis of gender differences in mathematics performance and consider the implications of the results for education, and then I review a meta-analysis of gender differences in mathematics attitudes and affect and consider its implications for education. Next, I share some observations on gender issues in mathematics and science education. A dilemma that is a continuing question throughout the chapter is this: Is it preferable to take a minimalist approach or a maximalist approach to

The research reported here was supported by the National Science Foundation, Grant MR 8709533. The opinions expressed are those of the author and not necessarily those of the National Science Foundation. Elizabeth Fennema was coinvestigator on the grant and provided innumerable insights into the nature of mathematics education and related gender issues.

gender differences in mathematics? That is, should scientists emphasize gender discrepancies and seek remediation, or should they emphasize gender similarities? The minimalist approach emphasizes similarities, whereas the maximalist approach emphasizes differences.

A Brief Introduction to Meta-Analysis

Meta-analysis provides a good summary of the voluminous research literature on gender and mathematics performance. Meta-analysis is a quantitative method for integrating the results of research from many studies. Essentially, it is a statistical method of doing a literature review.[1] In the meta-analyses on gender differences, the following formula is used:

$$d = \frac{M_M - F_F}{S},$$

in which d is the effect size; M_M is the mean for males and M_F is the mean for females; and S is the average within-gender standard deviation. Essentially, d measures how large the gender difference is by looking at the difference between male and female means, standardized by standard deviation units. For those readers who are unfamiliar with the d statistic, it can be compared with the familiar z score in that d, too, has a mean of 0, can take on positive and negative values, and has decreasing probabilities of scores very far from 0.

In performing a meta-analysis, the researcher first collects all available studies providing data on a particular question, in this case, gender differences in mathematics performance or ability. Care should be taken to locate both published and unpublished studies (e.g., dissertations) to guard against a possible publication bias toward studies that show significant gender differences. Computerized literature searches of databases, such as *PsycLIT* and *ERIC* (for unpublished studies), are invaluable in identifying as complete a sample as possible of available studies.

For each study providing relevant data (i.e., data on male compared with female performance on mathematics), the meta-analyst computes an effect size, d. At the same time, numerous characteristics of the study

---

[1] For a user-friendly introduction, see Hedges and Becker (1986).

can be coded, such as the year in which the study was published, the cognitive level of the mathematics test, and the age of the subjects. Multiple effect sizes can be obtained from a single article if data for multiple independent samples are presented (e.g., data for third, fifth, and seventh graders).

The meta-analyst can then compute a mean effect size, averaged over all studies, that estimates the magnitude of the gender difference in mathematics performance. Although there is controversy about interpreting these statistics, a general guideline proposed by Cohen (1969) is that a d of .20 is small, a d of .50 is moderate, and a d of .80 is large. The effect size d can also be related to the familiar correlation coefficient r, by the approximate formula $d = 2r$ (for large values of r, the exact formula should be used, $d = 2r/[1 - r^2]$).

Further statistical analyses have also been developed. One can test, for example, whether studies of a particular kind (e.g., samples of elementary school children) find a significantly larger gender difference than studies of another kind (e.g., samples of high school students). The technique of homogeneity analysis (Hedges & Becker, 1986) allows the investigator to explore variation (i.e., nonhomogeneity) in values of d across different kinds of studies.

Gender Differences in Mathematics Performance

In 1990, my colleagues and I published a meta-analysis of gender differences in mathematics performance (Hyde, Fennema, & Lamon, 1990). We located 100 studies that reported data on 254 different samples, representing the testing of more than 3,000,000 subjects. Averaged over all 254 samples, the mean d was $+.15$, a difference that is surprisingly small given psychologists' belief for decades that the gender difference in mathematics performance is a strong, reliable phenomenon. And, if we look only at samples of the general population (which are the best methodologically), d is $-.05$. That is, there is actually a small difference favoring females, although the difference is so small that it could be considered to be zero. To provide a visual representation of how large the gender difference is, Figure 1 shows two normal curves 0.15 SD apart.

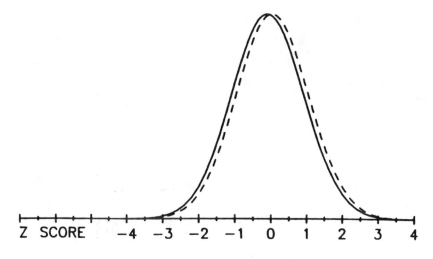

FIGURE 1. Two normal distributions that are 0.15 standard deviations apart (i.e., d = .15. This is the approximate magnitude of the gender difference in mathematics performance, averaging over all samples.) Reprinted by permission of the author.

I would argue, based on the small effect size of $+.15$ averaged over all samples, and the tiny but reversed effect size of $-.05$ based on the best, most general samples, that there are no overall gender differences in mathematics performance. That is, in this case, the results merit the conclusion of the null hypothesis. However, the data are more complex than that because the effect sizes are not homogeneous, so it is necessary to pursue additional analyses.

On the issue of sampling, we analyzed the data according to the selectivity of the sample, ranging from samples of the general population to moderately selective samples (e.g., college students), highly selective samples (e.g., Harvard students, graduate students), and highly precocious samples (e.g., Benbow & Stanley's [1980] SMPY project at Johns Hopkins University). The more selective the sample, the larger the gender difference. For samples of the general population, d is $-.05$; for moderately selective samples, d is .33; for highly selective samples, d is .54; and for precocious samples, d is .41. It is critical that we not generalize from highly selected samples to the general population. The pattern of results is quite different.

This phenomenon of larger gender differences among selective samples is explained, at least in part, as a simple statistical artifact somewhat like a restriction of range, causing reductions in the correlation between two variables. In this case, the highly selective samples produce a restriction of range in the mathematics performance variable. The result is a markedly reduced standard deviation in the numerator of the d statistic, which, in turn, inflates d.

A second explanation of this phenomenon must also be considered. The analyses reported here, and particularly the graph shown in Figure 1, assume that males and females are equally variable, that is, that the variance in mathematics performance for males is the same as that for females. This, of course, is an empirically testable hypothesis. The greater male variability hypothesis has appeared a number of times in the history of psychology (e.g., Maccoby & Jacklin, 1974). The hypothesis that men are more variable than women in some or all mental abilities is appealing because it can simultaneously explain the surplus of highly talented men and the surplus of men among the retarded, those with learning disabilities, and so forth. That is, it can explain the surplus of men at both ends of the distribution. Recent work by Feingold (1992) indicates that men are somewhat more variable than women on measures of mathematics performance. The ratio of the male to female variance is approximately 1.2:1 (see also Hedges & Friedman, 1993).

In another analysis, highly relevant to education, we examined the interaction between age and cognitive level of the test as it influences the magnitude of the gender difference. We coded all tests used in the studies into one of the following categories of cognitive level: computation, understanding of mathematics concepts, problem solving, mixed, or unreported. The results indicated a small female advantage in computation in elementary and middle school and no difference in high school, no gender difference in understanding of mathematics concepts at any age, and no gender difference in problem solving in elementary or middle school. The only place in which we found small-to-moderate differences favoring men was in problem solving in high school (d is .29) and in college-age samples (d is .32).

To summarize the results, the evidence indicates that, overall, there is no gender difference in mathematics performance. However, the results need to be stated in more complex form and this, in turn, raises the minimalist–maximalist dilemma. There are no gender differences in computation, in understanding of mathematics concepts at any age studied, or in problem solving in the early grades, but a moderate difference in problem solving emerges in high school. Should we take the minimalist approach and emphasize the conclusion of no gender difference (for which there is plenty of evidence), or should we take the maximalist approach and emphasize the difference in problem solving beginning in high school, which has implications for mathematics education? I hope that a complex, combined conclusion can be derived: that there are basically no gender differences in mathematics performance but that a moderate (not large) gender difference in problem solving begins in high school and needs attention.

Implications for Education

This complex message must be carried over when considering the implication for education. First, it is important to consider guidance counselors, who may have enormous influence on students' choices of courses. These counselors are surrounded by a media environment in the United States that emphasizes (indeed, glamorizes) the maximalist message (see, e.g., *Newsweek* [cover], May 28, 1990; New York Times [op. ed.], July 5, 1989; *U.S. News and World Report* [cover], August 8, 1988), although I have just explained that the meta-analyses, which provide by far the best scientific evidence on the point, contradict these maximalist assertions. These counselors need to hear the "no gender differences" message: that girls are just as competent at mathematics as boys and should be encouraged to have just as high aspirations for the courses that they take. Research indicates that a large part (although not all) of the gender gap in mathematics performance in high school is accounted for by the gender difference in mathematics course taking (e.g., Fennema & Sherman, 1977; Kimball, 1989). Rather than conduct further research on abilities, it may be better to focus on the issue of course taking, as Jacqueline Eccles and her colleagues have done so well (e.g., Eccles, 1987).

When communicating the gender similarities message, the difference in problem solving in high school and its implications for education must be considered simultaneously. Efforts are needed to make sure that girls continue to take mathematics courses every year in high school. But a related question considers where students learn mathematics problem solving in high school. Elizabeth Fennema (personal communication, June 1990) argued that, nationwide, mathematics problem solving is rarely taught and that the emphasis in most mathematics classes is almost exclusively on simple computation. In short, girls learn well what they are taught in the classroom, but fall behind in what they are not taught. Much experience in problem solving is gained in science classrooms (e.g., solving equations in chemistry). Girls then have a double disadvantage because they are both less likely to take mathematics courses in high school, and less likely to take science courses, thereby losing additional opportunities to learn problem solving.

From a public policy point of view, the tendency of high school girls to take fewer mathematics courses than boys might be approached in several ways. One would be to pass state legislation requiring 4 years of mathematics for all students in order to graduate. As the United States struggles to be competitive internationally in mathematics and science, such a public policy remedy seems not only reasonable but imperative. Short of this, universities, particularly state universities, would be well advised to have a basic admissions requirement of 4 years of high school mathematics.

The research of Bahrick (chapter 2, this volume) lends further urgency to the concern over the lesser mathematics course taking by women. In brief, his research shows that the more times one reviews the same course content, the better it is remembered, but also that forgetting occurs as time passes after the course. Consider the case of a woman who took only 2 years of mathematics, during her freshman and sophomore years in high school. Suppose that she is taking the SAT mathematics as a senior, or that she is attempting a mathematics course as a college freshman. Compare her to a man in the same situation, but who had a full 4 years of high school mathematics. The woman is disadvantaged in three ways: (a) She took only 2 years of mathematics and cannot be expected to do well on problems tapping knowledge from courses that she did not take. (b) It has been longer since her last mathematics course so that she has forgotten more. (c) She has missed reviews of mathematics material that occurred in the two courses that she did not take, creating further disadvantages in her memory of the material. It is not surprising, then, that she does not score as well on the SAT, or that she is somewhat apprehensive about taking a college mathematics course. The implication for education is again clear: We must make sure that all students take 4 years of high school mathematics.

This raises the issue of girl-friendly mathematics classrooms pioneered by Eccles, Fennema, Peterson, and others (e.g., Eccles & Blumenfeld, 1985; Peterson & Fennema, 1985). The notion is that some kinds of mathematics classroom organization seem to facilitate the performance and expectations of success of girls (e.g., cooperative learning structures) and others seem to facilitate the performance of boys (e.g., competitive learning structures). More research is needed on teaching styles that facilitate the mathematics and the science learning of girls, particularly in the problematic area of problem solving in senior high school. This research should not commit the maximalist error of assuming that girls and boys have completely different learning styles. Rather, it should recognize the great overlap in distributions and great within-gender variability. If it is found, for example, that girls do better, on average, when work is cooperatively organized rather than competitively organized, that does not mean that every girl does better under cooperative learning systems. The goal is to find the learning environment best suited to the individual. Gender may not be a very good guide for assigning students to the environment that is best for them. Research should focus on finding better indicators.

Gender and Mathematics Anxiety and Attitudes

In a second meta-analysis, we investigated gender differences in mathematics attitudes and affect (Hyde, Fennema, Ryan, Frost, & Hopp, 1990). We located 70 usable articles, reporting on 126 separate samples representing the testing of 63,229 subjects. We looked at all the constructs covered by the much-used Fennema–Sherman scales: mathematics self-confidence, mathematics anxiety, usefulness of mathematics, stereotyping of mathematics as a male domain, attitude toward success in mathematics, sense of mathematics effectance, mother's attitude, father's attitude, and teacher's attitude. We also found measures of attributions of success and of failure in mathematics, and expectations of success in mathematics. The number of variables is so great that I do not review the results for all of them but instead select three constructs that are especially important.

Again, the popular media have saturated the public with views about mathematics anxiety: It cripples one for work in mathematics, turning normal, rational people into sniffling, ineffectual idiots; and women are far more likely to be afflicted with this problem than men are. Indeed, some see mathematics anxiety and its alleviation as a women's issue. What do the data say?

Averaged over all samples, d is $+.15$ for mathematics anxiety, indicating that women report (note that these are self-report measures) more mathematics anxiety, but the difference is small. The data lead to a minimalist approach: There is not much of a gender difference in mathematics anxiety. Mathematics anxiety may still be an important issue, one that we should work on, but the data do not indicate that it is a gender issue.

We also had a good deal of data on mathematics self-concept or self-confidence. Here d is $+.16$, again indicating a difference favoring men but a small one. Women are not overwhelmingly burdened with lack of confidence in their ability to do mathematics. A minimalist conclusion seems appropriate here.

One might pose the developmental question of the data: Are there age trends in gender differences in mathematics self-confidence? Might the small effect size of .16 be the result of averaging differences of 0 in elementary school, for example, with larger differences in high school? The answer to these questions is complex. Unfortunately, few studies have examined the mathematics self-confidence of elementary school children. We located none using the Fennema–Sherman scales and only four using other scales. Those found that d is $+.08$. Among high school students, d is $+.25$, a somewhat larger difference than among younger age groups, nonetheless a difference that would be classified as small.

One of the Fennema–Sherman scales was interesting because the pattern of results was considerably different. For the scale that measures the extent of the respondent's stereotyping of mathematics as a male domain, d is $-.90$. This is an enormous gender difference, larger than any other discussed in this chapter. It indicates that males stereotype mathematics as a male domain considerably more than females do. Although there has been much talk about how girls' stereotypes of mathematics as masculine might discourage them from pursuing mathematics courses or mathematics careers, the evidence does not support this view. Rather, it is men who are more likely to stereotype mathematics as masculine. Therefore, we might focus more profitably on significant men in the system (male mathematics teachers or male guidance counselors) to determine whether their attitudes have an influence on girls and on their pursuit of mathematics.

Observations on Gender Issues in Teaching Mathematics and Science at the University Level

In my administrative role as associate vice chancellor at the University of Wisconsin, I was charged with advancing gender equity for women faculty, staff, and students. Therefore, I have worked a great deal with gender issues in the sciences, mathematics, and engineering, and can share some of the observations that I have made about what hinders the educational progress of women in these areas. I do so because I think psychologists sometimes run the risk of trying to explain the dearth of women among professors of physics by focusing on complex information-processing models while ignoring more obvious factors that have great explanatory power. What are some of these factors?

1. *Lack of role models.* Among faculty at prestigious research universities, less than 10% of the faculty in mathematics and the sciences are women. Indeed, at the University of Wisconsin, many science departments have no women on the faculty. It is not surprising, then, that women students do not choose to pursue occupations (and the courses

that lead up to them) in which they see no women. Vigorous programs are needed to hire women faculty members, as well as vigorous efforts to "prime the pump" by enlarging the number of women graduate students earning doctoral degrees.

2. *Overt discrimination including sexual harassment.* In a recent case, a talented young woman from South America came to work as a graduate student in the laboratory of a male scientist at a major research university. When she arrived, she discovered that he expected her to form a romantic relationship with him and, if that went well, to marry him. When she did not comply, he evicted her from the laboratory and cut off her funding. When the case was investigated, it was discovered that this was the second woman whom he had treated in this manner. Women cannot succeed in the sciences with this kind of sexual harassment. The psychological research that is much needed is a systematic survey of graduate student women (and perhaps undergraduates as well) majoring in the sciences and engineering to determine the incidence of harassment and other discrimination and the nature of these problems. From this can flow policies in which professionals police themselves successfully.

Summary and Conclusion

In summary, meta-analyses show that there are no gender differences in mathematics performance except in complex problem solving beginning in high school, and the issue there may be not lack of ability but failure to continue to take mathematics courses. The analysis using the effect size, d, indicates that gender differences in mathematics performance are small overall. Stated another way, within-gender variation is far greater than between-gender variation. In terms of research such as Siegler's (chapter 11, this volume) that examines styles of problem-solving, variation probably translates into a variety of problem-solving strategies among women and an equally great variety among men. Gender, therefore, is not a good predictor of mathematics performance or problem-solving strategies. Although researchers should continue to test for gender differences as new studies on students' learning processes emerge, I predict that few differences will be found and that they will be small in magnitude.

Gender is unlikely to be a good basis for assigning students to different methods of instruction.

Gender differences in mathematics anxiety and self-confidence are surprisingly small and close to zero. A minimalist approach seems to be supported by the data. On the other hand, a large gender difference is found in the stereotyping of mathematics as a male domain, with men holding much stronger stereotypes than women do.

Finally, while engaging in sophisticated research on gender differences in mathematics performance and attitudes, one should not lose sight of obvious factors that are barriers to women's entrance into careers in mathematics and science, including the lack of female role models and various forms of sex discrimination, including sexual harassment. Here, too, psychologists can contribute helpful research.

References

Benbow, C. P., & Stanley, J. C. (1980). Sex differences in mathematical ability: Fact or artifact? *Science, 210*, 1262–1264.

Cohen, J. (1969). *Statistical power analysis for the behavioral sciences.* San Diego, CA: Academic Press.

Eccles, J. S. (1987). Gender roles and women's achievement-related decisions. *Psychology of Women Quarterly, 11*, 135–172.

Eccles, J. S., & Blumenfeld, P. (1985). Classroom experiences and student gender: Are there differences and do they matter? In L. C. Wilkinson & C. B. Marrett (Eds.), *Gender influences in the classroom* (pp. 79–114). San Diego, CA: Academic Press.

Feingold, A. (1992). Sex differences in variability in intellectual abilities: A new look at an old controversy. *Review of Educational Research, 62*, 61–84.

Fennema, E., & Sherman, J. (1977). Sex related differences in mathematics achievement, spatial visualization, and affective factors. *American Educational Research Journal, 14*, 51–71.

Hedges, L. V., & Becker, B. J. (1986). Statistical methods in the meta-analysis of research on gender differences. In J. S. Hyde & M. C. Linn (Eds.), *The psychology of gender: Advances through meta-analysis* (pp. 14–50). Baltimore: Johns Hopkins University Press.

Hedges, L. V., & Friedman, L. (1993). Gender differences in variability in intellectual abilities: A reanalysis of Feingold's results. *Review of Educational Research, 63*, 94–105.

Hyde, J. S., Fennema, E., & Lamon, S. J. (1990). Gender differences in mathematics performance: A meta-analysis. *Psychological Bulletin, 107*, 139–155.

Hyde, J. S., Fennema, E., Ryan, M., Frost, L. A., & Hopp, C. (1990). Gender comparisons of mathematics attitudes and affect: A meta-analysis. *Psychology of Women Quarterly, 14*, 299–324.

Kimball, M. M. (1989). A new perspective on women's math achievement. *Psychological Bulletin, 105*, 198–214.

Maccoby, E. E., & Jacklin, C. N. (1974). *The psychology of sex differences.* Stanford, CA: Stanford University Press.

Peterson, P. L., & Fennema, E. (1985). Effective teaching, student engagement in classroom activities, and sex-related differences in learning mathematics. *American Educational Research Journal, 22*, 309–335.

CHAPTER 8

Gender Differences in Cognitive Style: Implications for Mathematics Performance and Achievement

Diane McGuinness

*C*ognitive style refers to a consistent strategy choice for problem solving and organizing information for later retrieval. Children work at making sense of their environment and in adapting this understanding to tasks set for them by parents and teachers. In each experience and in every task, there are multiple sense impressions and options for potential strategies. Objects (animate or inanimate) have color, form, size, roughness, heaviness, and functions, as well as names. What a child chooses to notice about objects and events depends on what sensory cues are salient for that individual child. Even animals of the same species show individual differences in what they observe and ignore about a stimulus (Trabasso & Bower, 1968).

Unless revealed by experiment, cognitive style is often part of tacit knowledge, typically invisible both to the adult observer and to the child. When a child finds a solution to a problem, it may not be apparent to the child precisely how he or she arrived at that solution. This is especially so in complex learning tasks like reading and mathematics. To pry open this inner world, experiments must be framed to discover how many strategies are available for any given task and which are most efficient.

There are two major approaches to the study of cognitive style. One is to make assumptions about the range of potential strategies that could be adopted to solve a particular problem and to test these assumptions on a random population. Another is to classify individuals along certain dimensions and to look for differences among groups. Age is an obvious category, as is gender. Studying individual differences leads to an analytic approach that is derivative rather than presumptive. Contrasts among individual groups in performance on identical tasks focuses the analysis on the reasons for those differences, reasons that may never have occurred beforehand. The Piagetian framework for developmental psychology was built on this type of approach.

Gender is an important category for the study of mathematical strategies because of the highly skewed distributions in tests of advanced mathematics. Men are overrepresented in high-ability groups and in occupations requiring advanced mathematics skills. These gender differences cannot be written off as accidental or caused by sociocultural factors because they appear in rudimentary form by about 4 years of age. From the evidence to date, boys are more likely than girls to generate a three- or even a four-dimensional mental model of the spatial organization of the world of objects and object relations. Girls are more likely to take a linguistic approach to most problem-solving tasks and to rely more heavily on verbal memory. The theoretical position of this chapter is that these strategy biases have important implications for the development of a knowledge system and that they impact most profoundly on performance only when a particular strategy is much more effective than any other.

Although there may be a biological basis for these sex differences, this does not mean that biology is destiny and that sex differences in learning style are immutable or immune to training. Proper pedagogy and timely intervention could result in *both* sexes achieving a much greater potential than is currently the case. Instead of ignoring the sex differences and pretending that they do not exist, they could be used to provide much needed evidence on the strategies that succeed or fail in certain types of mathematical performance.

Before taking up the evidence in greater detail, it is necessary to deal with the social models put forward to explain sex differences in mathematics. Unfortunately, sex difference research has been overly politicized. Because of this, a chapter on sex differences in mathematics problem solving must begin with an analysis of social models put forward to explain women's underrepresentation in programs of advanced mathematics and the "hard" physical sciences, such as engineering and physics.

The basic facts are these. In elementary school, girls and boys perform similarly on standardized achievement tests. By the time more abstract mathematics concepts are introduced into the curriculum, usually toward the end of junior high school, girls who were previously doing well start to fall behind. By the time advanced examinations like the SAT enter the picture, girls perform, on average, 47 points below boys on the mathematics portion of the test (College Entrance Examination Board, 1987). This relatively poor showing blocks many young women out of programs in mathematics and science at the college level, closing off many career opportunities.

Sociocultural Theories

The Social Conspiracy Model

The most extreme theory put forward to explain this state of affairs is that women fall behind in mathematics performance because of sex discrimination (the *social conspiracy model*). In this model, adults in and out of the classroom conspire to create a bad attitude in girls toward mathematics (even math phobia), causing them to give up and drop out of mathematics classes. Even when they do not, this theory carries the additional corollary that attitudes cause mathematics performance, so that any sign of poorer mathematics skills in girls enrolled in high school mathematics courses is attributed to attitudes rather than to ability.

This model has had little support from the vast amount of research devoted to it (see Benbow, 1988, for a lengthy review of these studies). Furthermore, it fails on logical grounds alone. Why does the conspiracy against girls start so late? Why is it more specific to geometry than to algebra? Why does this conspiracy not exist at all levels of the system for all subjects? Why is there no conspiracy against girls in history and geography, or in the life sciences like biology, botany, physiology, and zoology in which girls do well? The fact that this theory has been so predominant is unfortunate, because we have wasted an enormous amount of time and resources trying to bend science to a political philosophy, instead of finding out what is really happening.

There is a further difficulty with the social conspiracy model in that it contains no corrective principles other than for society to rearrange its stereotypes. The fact that this is very difficult to accomplish and extremely slow is attested to by a comparison of SAT mathematics scores over the period 1967 to 1987. During this time, there were continual efforts on the part of feminists to raise consciousness about sex discrimination and to persist in monitoring and eliminating sexist content in textbooks and in national achievement tests like the SAT. Despite these efforts, over this 20-year period, women's SAT mathematics scores relative to those of men remained unchanged at -47 points (College Entrance Examination Board, 1987).

Thus, both the data from studies based on the social conspiracy model and the efforts on behalf of women show the theory in error and the social experiment a failure, at least for improving women's mathematics scores or for enticing women into the physical sciences. Clearly, something better is needed. As everyone agrees, if women are being closed out of important careers in technology and science because of problems with higher mathematics, then a great deal of energy and talent may be wasted. Failure to acquire one's academic potential closes off career opportunities and deprives the nation of much needed expertise.

Before exploring the alternative hypotheses, it is important to disentangle two factors that have been confounded in this overly political approach.

First, there is undoubtedly discrimination against women in the job market. Study after study attests to the lower levels of pay for women with comparable training and experience to men. Surveys continually show that women fail to attain positions of authority and status that would be commensurate with their ability, in a variety of fields from politics to business to academia (see Gallagher, chapter 1). Brush (1991) reviewed surveys that show that women are at a salary disadvantage in all disciplines of mathematics and science, even in those areas in which women are most numerous, such as psychology, biology, the social sciences, and medicine.

However, this is an entirely different issue from ability or talent. Sex differences show up repeatedly in tests of advanced mathematics, even in those cases in which boys and girls have had an identical number of courses. Benbow and Stanley (1983), in their study of precocious youth at junior high school age, concluded that the children who performed at very high levels on the SAT (a high-school-level test) were showing obvious talent far and above what could possibly be predicted from their formal training in school. The sex ratios (male:female) became more exaggerated the higher the score: At 700 or higher ($n = 280$), the sex ratio was 13:1. This is a very powerful argument for highly skewed distributions for mathematics aptitude between males and females, and especially noteworthy as all the girls in the study claimed to like mathematics and obtained good grades in those classes. Whereas some may view this finding as discomforting, we cannot afford to replace science with emotion. Instead, we need to look at other reasons that might predict these results.

Only a few people are truly mathematically gifted. Most people who go into the physical sciences and need mathematics skills at a very high level develop this ability with training. Thus, ability (performance) can be independent of giftedness or talent. Talents may or may not be developed or exercised. This means that there are additional and complex issues in trying to identify what is holding women back from careers in mathematics and science. Do they have less innate talent than men, less ability than men, or both, or neither? This is a critical issue because ability is trainable (modifiable), whereas talent is biological (innate). Mathematics expert Pat Davidson, who has done extensive work on how to teach mathematics skills, maintains that everyone can learn mathematics to a very high level of expertise (McGuinness, 1985). Although there will always be those who are truly gifted, as in any other field, most people could do infinitely better than they do, if they had proper teaching. The real question in the search for the reasons for sex differences in mathematics aptitude is why boys can acquire useful skills from the same bad teaching that later give them an edge in mathematics performance.

The Cultural Accident Theory

As noted earlier, in achievement test performance gender differences do not appear on general tests of mathematics until quite late, when mathematics becomes more abstract and complex. There are problems with achievement tests, which is outlined shortly, but the basic findings from global surveys such as those of Hyde, Fennema, and Lamon (1990) and reported by Hyde (chapter 7) is that gender differences become more noticeable with age and with the difficulty of the test, and are most likely to appear in higher level tests like the SAT, the Graduate Record Examination (GRE), and the Graduate Management Test, or in very-high-ability groups like those tested by Benbow and Stanley (1983). Hyde et al. (1990) reported effect size differences that varied from around .40 on the SAT to .67–.77 on the GRE. The actual proportional difference between the sexes can be computed by the binomial effect size (effect size 2); thus, GRE effect sizes of .67 compute to a 33% higher performance in men. This is certainly large enough to block many women from entry to graduate programs requiring mathematical competence.

On the other hand, results from tests taken in elementary school may even show a slight advantage for girls. The apparent late appearance of gender differences could be due to a variety of factors. One suggested by Hyde et al. (1990) could be called the *cultural accident theory* in which mathematics is stereotyped by our culture as "male," encouraging more boys to continue with advanced mathematics and discouraging girls. This theory and the social conspiracy model both ignore the possibility that there may be differential degrees of aptitude in highly specific content areas. These subtle differences may be lost when vast amounts of individual test items are averaged together in achievement tests like those reported in the Hyde et al. survey. Indeed, these authors commented on the fact that most of the studies that they reviewed failed to report any information on test content.

Both hypotheses are testable. The first predicts cultural variation between the sexes in mathematics performance in which sex differences either do not appear at all or go in the reverse direction from those in the United States. The second hypothesis predicts that mathematics expertise might appear earlier in boys if specific content areas were measured carefully.

There are several studies that fulfill both of these requirements. The most extensive, reported by Lummis and Stevenson (1990), involved three cultures (Taiwan, Japan, and the United States) and testing at kindergarten, first, and fifth grades. Using 2,000 children per country and group tests, the authors found no effects of gender, but large effects for country (Asians superior) were found when the data were computed as one standard score for the mathematics test as a whole. This finding confirms the results of many studies (noted previously) that have shown essentially no gender difference in general mathematics performance at elementary school.

In a second part of this same study, the authors used a smaller sample ($n = 720$), at first and fifth grade, to test each child individually on a variety of mathematics content areas. These included word problems, concept of number, mathematics operations, reading graphs and tables, measurement, estimation, computational speed, and two tests of spatial visualization. The same sex differences for all three countries were found on several subtests. Boys were superior at first grade on word problems ($p < .05$), estimation ($p < .01$), and visualization ($p < .05$). At fifth grade, results were similar, with boys superior on word problems ($p < .01$), measurement ($p < .05$), estimation ($p < .05$), and visualization ($p < .01$). In general cognitive tests at fifth grade, girls were found to be superior on auditory and verbal memory, and on coding speed; boys were superior on spatial relations tests and on general information. There were no significant interactions on any test between sex and culture.

In comparing the attitudes of the children toward the various subjects in school with their actual performance on tests, essentially no relationship was found. Thus, if parents perceive mathematics, reading, and so forth as essentially sex-specific domains, these attitudes do not influence their child's test performance. Instead, it was found that a mother's predictions of an individual child's ability correlated with his or her performance.

Other cross-cultural studies, all using high school students, show that boys are superior in mathematics in general tests in countries such as Great Britain (Harvey & Goldstein, 1985), Australia (Pattison & Grieve, 1984), New Zealand (O'Halloran, 1983), and Israel (Zeidner, 1986). The findings from these studies on the analyses of individual content areas are consistent in showing that boys are particularly advantaged in spatial problem solving or spatial reasoning, mechanical reasoning, and problems dealing with proportionality or scale.

It would be of considerable scientific interest to discover whether the particular tests that showed sex differences in the Orient and the United States, those involving estimation, visualization, and the ability to solve word problems, are independent constructs or part of a general factor that reflects the ability to visualize object relations and motion. In other words, are these subtests part of a general spatial ability? In view of the strong support for sex differences in spatial ability, this seems likely. If so, Lummis and Stephenson's (1990) results would provide evidence for the emergence of sex differences in spatial ability by the age of 6 years.

Whatever proves to be the case, if there is a stereotype of mathematics as a male domain, this may be because mathematics is a male domain. Males are better in certain aspects of mathematics than females. What appears to be a slight advantage in early spatial reasoning could become a bigger advantage whenever spatial visualization is a crucial component of the problem-solving process, as, for example, in geometry.

Sex Differences in Cognitive Style

An alternative hypothesis to the environmental theories outlined earlier is that the sexes adopt different strategies or styles in learning and in applying mathematics concepts. To understand how this might happen, it is important to discuss what mathematics is actually about.

What Is Mathematics?

Mathematics, like any symbol system, is a representation of concrete objects or sensory events in an arbitrary form: a set of symbols. The most elementary form is a counting system, or a fixed sequential ordering of individual symbols (numbers) that represent an increase in quantity. To understand a counting system, however, one must also understand the principles by which concrete objects can be ordered: They can be increased or diminished serially or in multiples. Also, counting relies on number systems using repetitive units or bases (base 2, base 3, and our system: base 10). Because much of this information is too abstract to be taught to young children, the best possible solution is to allow the child to encounter physical objects that can be ordered along fixed or limited dimensions, and to connect a number system to this concrete representation. Unfortunately, this is not usually the way in which early mathematics is taught. Often a child learns to move backward and forward in a counting system with little understanding of what the numbers represent. Every child, unless specifically instructed to make connections between objects, pictures, and symbols, unwittingly adopts a strategy to "make sense" of this enterprise. The teacher is generally unaware of this strategy because problems can be solved at this stage in various ways. The most appropriate has the richest set of representations: a mental landscape of objects, object properties, object relationships, and a sense of ordering along dimensions of quantity, size, shape, and so on. The most inefficient is the memorization of specific algorithms, which allows the child to get the "right answer" by memorizing a counting system and the rules of place value, and by carrying numbers across from one place to the next. As there is abundant evidence that boys are more interested in objects and object properties, and that girls have excellent verbal and pictorial memories (Maccoby & Jacklin, 1974; McGuinness, 1985; McGuinness & McLaughlin, 1982; McGuinness, Olson, & Chaplin, 1990), it may not be surprising that, when left to grasp at straws, the child falls back on his or her most efficient learning style. Boys connect symbols (numbers) to objects, but go astray in setting them out neatly on worksheets. Girls fail to connect symbols to much of anything, but are maximally efficient in managing a counting system and memorizing algorithms, and are especially good on paper because of their early superiority in fine-motor control (McGuinness, 1985).

These strategies are valid for a long period of time and work well for partial numbers such as fractions, percentages, and decimals. They remain valid through beginning algebra, which is often taught strictly according to fixed algorithms: "The denominator on one side of the equation becomes the multiplier on the other side of the equation." (Students are almost never told what algebra is for, even when they ask.) But during this period, somewhere in the transition from prealgebra to algebra, to geometry, and to calculus, individual learning styles begin to matter. Having a concrete mental representation to add to an algorithm "makes sense" of the process. As algebra becomes more abstract, it is actually a recoding of the more linear arithmetic principles into a two-dimensional spatial arrangement. New domains enter into the process, such as solving for unknowns. New symbols, borrowed from the alphabet, stand for characteristics of numerical values. The spatial organization of known and unknown quantities on the page represent object or event relationships up to and including linear and quadratic functions. Finally, in geometry and in much of the mathematics that follows, the importance of spatial reasoning and the ability to conjure up mental landscapes is of paramount importance.

Spatial Thinking as a Cognitive Style

In her work with children and young adults of all ability levels, Davidson (see McGuinness, 1985, for a detailed account of her findings) discovered two major types of problem solvers. Type 1 is able to memorize a counting system and to move back and forth in this counting system by applying memorized algorithms. These children are very poor at estimating, and often do not recognize when an answer is completely wide of the mark. Type 2 is a holistic thinker, anchors processing in the world of objects, rarely gets an impossible answer, but becomes impatient with multistep procedures. These are the children who make "silly" mistakes on worksheets and become bored with repetition.

Davidson has not assessed which of the two types is most likely to be male or female, but from her years of experience in teaching college mathematics, she notes that by the time women get to her classes, they are noticeably lacking in skills in spatial visualization and estimation, and in any understanding of the concrete properties of abstract mathematics (McGuiness, 1985).

Lummis and Stephenson's (1990) description of the areas in which men were superior (visualization, estimation, measurement, and word problems) is consistent with the hypotheses about male and female mathematics styles outlined earlier. According to Lummis and Stephenson's description, word problems also benefited from visualization skills. One of their examples went as follows: "There were 156 boxes of oranges that had to be delivered to a store. The truck held 33 boxes. How many trips did the truck have to make?" As there is no clue concerning the algorithm in this example, the way to solve the problem is to visualize 156 boxes standing on the pavement or in a loading bay, with the empty truck waiting to be filled with the first 33 boxes. One solution is to subtract 33 from 156 and to continue subtracting by 33s until there is a number equal to or less than 33 remaining. The other solution is to recognize that this tedious process can be converted into a problem of long division. Either solution will work, but both require concrete imagery of the scene and of the movement of objects between locations.

Lummis and Stephenson's (1990) data, as well as the cross-cultural research cited earlier, strongly implicate a major distinction between boys and girls in the use of visual or visuospatial strategies in mathematical reasoning. If this is so, when and under what circumstances would this difference begin?

In a study on spatial reasoning in preschoolers conducted by McGuinness and Morley (1991), 137 children aged 3 to 5 years were presented with two types of puzzles: two-dimensional jigsaw puzzles, and a three-dimensional problem constructed out of Lego blocks. Children were tested individually and told to complete the puzzles as quickly and as accurately as possible. They were timed with a stopwatch. No sex differences were observed in any of the jigsaw puzzles in terms of speed or accuracy. However, when children were asked to build a replica of a three-dimensional Lego block pattern, the boys were significantly faster than the girls from the age of 4 years on. The sexes did not differ in error scores.

In preparing this chapter, I went over recent data on a study designed to measure gender differences in preschool children in behavioral tempo (McGuinness, 1990). In that study, I looked in detail at everything the child did within a certain time frame, focusing specifically on number of tasks attempted, interruptions, and time on task, rather than on an analysis of the play material per se. However, I did code separate types of play, such as constructional play and teacher-organized play. Girls were more likely to engage in teacher-organized play (largely crafts and painting) and boys in constructional play. However, a number of girls did play with materials classified as constructional. When reexamining the data, I focused on the type of constructional play in more detail. As a general conclusion from this exercise, boys were found typically to add a third or even fourth dimension to play with objects. Three-dimensional block constructions were common. When the boys built a two-dimensional surface, it was usually a surface to convey moving objects (two dimensional + motion). Some boys added a fourth dimension, three-dimensional structures with embedded conveyors: bridges, slopes, and passages. Girls' constructions were most often two-dimensional, static patterns using color. They spent most of their time working with pattern blocks, mazes, mosaics, and jigsaw puzzles.

Similar findings were observed by Jahoda (1979) in a cross-cultural comparison among children aged 7, 9, and 11 years from Scotland and from Ghana. Boys from both countries were superior in constructing a replica or a three-dimensional block pattern, but performed equally to the girls on two-dimensional problems. Ghanaian children were dramatically disadvantaged when the problems were transformed into two-dimensional pictorial representations, although they performed similarly to the Scottish children when the tasks were three dimensional. This suggests that three-dimensional problems represented by two-dimensional line drawings, like those most commonly found in spatial reasoning tests, are highly abstract and may not be a true reflection of spatial ability.

Nevertheless, one of the most consistent sex differences yet observed is the male advantage on tests of spatial reasoning. These data have been reviewed elsewhere and are not presented in any detail here (see Harris, 1978; Maccoby & Jacklin, 1974; McGuinness, 1985). What is relevant in addressing this body of research is that in all cases in which a sex difference is found, the task requires the image of objects or of parts of objects *in motion*.

This is true whether the drawing is a line drawing of a complex unfolded three-dimensional shape that must be folded in the mind (Differential Aptitude Test Spatial Relations subtest); rotating shapes represented pictorially; or Piaget's "infamous" water-level test, in which women do notoriously poorly. In this task, the subject must estimate the angle of the water line on a drawing of a tilted pitcher. To solve this problem correctly, the subject must make an inference about the relationship between the water and the pull of gravity. Liben and Golbeck (1984) tested a large number of college students on this and on a similar task in which the subjects had to draw a cord and a light bulb on the car of a train shown climbing a steep gradient. They found that men were noticeably superior in both tasks and that only 50% of women could use information about the pull of gravity even when instructed to do so.

Perhaps the most definitive study on sex differences in pencil-and-paper tests of spatial ability was the attempt by Barrett (1955) to replicate Thurstone's (1950) original findings for a three-factor solution for spatial reasoning. Barrett used 10 different tests for spatial reasoning, and factor analyzed the data independently by gender. He found that men were statistically superior on 9 of the 10 tests. He was able to replicate Thurstone's original factor structure for the men only. Factor 1 was the ability to recognize objects seen from different angles. This involves imagining the three-dimensional properties of a rigid object and its motion around an axis. Factor 2 was the ability to imagine movement or displacement among the various parts of a static object. Factor 3 was the ability to imagine that one had moved around an object and was viewing it from a different location in space. All 10 tests loaded on one general factor for the women.

Artificial tasks like the ones just described may or may not be relative to performance in real-world settings. However, these tasks have been shown, in a number of studies, to be highly correlated with certain types of higher mathematics, most especially plane and analytic geometry (Barakat, 1951; Fennema & Sherman, 1977; McGuinness, 1985; Stallings, 1979). Greater facility in spatial reasoning could be expected to provide a boost in mathematics performance, particularly at advanced levels of the discipline.

Very few studies have been conducted on sex differences in real-world situations in which spatial reasoning would be expected to play a part. Two studies by McGuinness and Sparks (1983) looked at the differences between men and women in their representations of a familiar terrain: the college campus at the University of California, Santa Cruz. In the first study, subjects had to draw a map for a visiting friend, and in the second, subjects had to draw all connectors (roads, paths, and bridges) among a set of centrally located buildings drawn to scale. Men adopted a strategy in which they framed space using geometric coordinates and the road system. They were considerably more accurate in placing buildings with regard to radial geometric coordinates referenced to a centrally located building, and represented significantly more paths and roads in their maps. They were also more accurate in drawing connectors in the second study. The women drew many more buildings in their maps, and were somewhat better at drawing to scale. McGuinness and Morley (1991) concluded that the men were more likely to frame space as a geometric layout of buildings and roads, whereas the women were more likely to frame space according to proximity (near to or far away).

The ability to conjure up mental representations of three-dimensional objects and to imagine them in motion is more than just a bonus for the mathematical mind. In certain cases, it is the essence of the process, the sine qua non without which the symbols in the formula would have no meaning. This ability must provide an advantage not only for performance on tests of mathematics, but also in the capacity to imagine solutions for problems in sciences like physics and engineering.

The mystery in these findings is how does one come to "get" such a strategy? Is this capacity for anchoring mathematics in the concrete world of objects locked into brain cells, is it a property of genes, or is it merely a different way of looking at the world, available to all of us, if we would look at the right time and in the right direction? Does the failure of some children to acquire the right set of mental codes for solving mathematics problems reflect on their inherent lack of talent, or on the teacher's inability to point out what is relevant?

Sex Differences in Interest: Objects Versus Persons

One additional hypothesis needs to be explored before we try to answer these questions. The problem of individual differences in cue salience, with which I began this chapter, takes on new significance when men and women are concerned. Because we are bombarded by an enormous number of stimuli at any moment in time and cannot attend to all of them, we pay attention instead to what is relevant in our lives—to what we are interested in. There is now a body of data that indicates that women are more interested in animate objects (especially people) than inanimate objects and that men are interested in inanimate objects more than in people.

Sex differences in this object–person domain were first demonstrated by Goodenough in 1957. She presented abstract mosaic patterns to young children and asked each of them to make up a story about the pattern: Eighty percent of the 2-year-old girls told stories about people versus 10% of the boys. By age 3, 90% of the girls talked of people versus 40% of the boys; and by age 4, 100% of the girls talked about people as opposed to 62% of the boys. Boys talked instead about things like trains, cars, balls, and various objects in their lives. Feshbach and Hoffman (1978) asked young children to describe events or situations in their lives that made them happy, angry, or sad. For most of the girls, other people's behavior (especially parents') was the cause of their remembered emotions; whereas for the boys, objects were as likely as people to be the source of feelings.

In a study by McGuinness and Symonds (1977), using the technique of binocular rivalry, college students had to report their perceptions when viewing pairs of slides in a divided visual field, one to each eye. The pairs consisted of pictures of persons paired with a picture of a common object such as a wristwatch or an automobile. Subjects were to report exactly what they saw at fixed intervals of time. In this situation, the most interesting (subjectively) of the two pictures suppresses the least interesting. There was a significant interaction ($p < .001$) between sex and the type of slide. Men preferred objects to people and at a higher rate than did women ("objects" were reported most frequently). Women were exactly the reverse, preferring people to objects.

In a study on visual memory for pictures (colored slides), college students were tested in both recognition memory and in providing a written description of the contents of the slide (McGuinness & Mc-

Laughlin, 1982). Men and women performed identically on the recognition part of the test, but women remembered significantly more content during written recall. Their superiority was due largely to their ability to remember the people in the slides.

So far, the data indicate that there may be two factors operating against women in achieving expertise in higher mathematics and in the physical sciences. The first is that more girls than boys may unwittingly adopt an inefficient strategy when learning basic mathematics skills in the elementary grades. The consequences of this strategy choice is that girls learn to manipulate a counting system rather than mathematics concepts, and do not connect this counting system effectively either to a number system or to concrete representations. This strategy will work (produce correct answers on work sheets and examinations) for an extended period of time, especially if the child has an excellent verbal symbolic memory.

Later, when mathematics concepts become more abstract and the symbolic formulations less transparent and "arithmetic," girls start to have difficulty. Boys, whose orientation is more likely to be visually based and more concrete, can use mental schema to make sense of the symbolic formulations. In addition, their sensitivity to multiple spatial dimensions makes geometric formulas relevant to something. For many boys, mathematics has meaning because it is experienced as a representation of something real.

The second problem is that girls are not as interested as boys in the domain to which mathematics is referenced: the world of objects. Mathematics is not about people or about anything living. One cannot do fractions, decimals, or percentages on dogs or horses, find the area of a baby, or do linear equations on parakeets and kittens. Mathematics, as the late Bronowski (1973) pointed out, is about a world that can be reordered. He made a distinction between the hand that molds nature (traditional cultures) and the hand that splits nature. Mathematics is about splitting, taking things apart and reorganizing the object parts in space, or describing paths of motion or objects through space. One can create a parabola by firing a cannonball, but not by firing tigers or trees.

State of the Art: Classrooms and Learning Styles

Reading as a Case in Point

The typical school classroom is filled with children who are guessing their way to an education. Education in America is, in essence, a lottery. Children have to figure out what strategies to apply to specific areas of learning. Often the strategies that they adopt are not only incorrect but actually counterproductive. Many children (maybe as many as 50%) come to believe that the way to learn to read is to memorize visual patterns or letter sequences, one unique pattern for each word. This strategy is so memory intensive that it begins to break down by about third or fourth grade. Not only are most teachers unaware that this is going on, but they inadvertently contrive to promote it. "Look–say" and "whole language" approaches, as well as early emphasis on "sight words," create inefficient readers.

The reason for this state of affairs is that until recently, research in education was largely the province of university departments of education. They have produced more than 50 years' worth of research on the pedagogy of reading, spawning tens of thousands of research reports. Unfortunately, much of this research is useless. Methodology is poor and haphazard, hypotheses follow fads instead of scientific principles, multivariate analyses are rare, and the results are frequently contradictory. As a consequence, educators responsible for writing textbooks and for training teachers have come to distrust research and have given up trying to make any sense out of it. Instead, teaching is basically driven by unsubstantiated fads, and teachers lurch from one fad to the next at the dictates of the school principal, the school board, and the textbook publishing houses.

It is not presumptuous to say that the major hope for education will come through cognitive psychology. The problem will be in getting the educational community to pay attention to the research findings. Already cognitive psychology has made an incredible contribution to reading research. Over the past 10 years, there has been a revolution in discovering what a reader needs to know to use a phonetic alphabet efficiently, a revolution brought about entirely through scientists working in the tradition and framework of cognitive psychology (Bradley & Bryant, 1983; Goswami & Bryant, 1990; Liberman & Mann, 1981; Rayner & Pollatsek, 1989; Wagner & Torgensen, 1987). But it may take a long time for this information to reach the classroom.

What About Mathematics?

Research on reading is light years ahead of research on mathematics, for which the fundamental questions have barely been addressed: What is it that a child needs to know to be an efficient problem solver in mathematics? How many kinds of mathematics are there? Does each require a different set of aptitudes or strategies? How do these strategies overlap? Are multiple strategies better than single strategies?

Earlier it was pointed out that the mathematics content areas investigated by Lummis and Stephenson (1990) were not submitted to factor analysis to determine any commonality between them. The most outstanding and comprehensive mathematics achievement test for elementary school-age children, the Key Math Diagnostic battery, has 13 subtests. But are these subtests really separate mathematics domains or subjective domains spawned out of the mind and experience of the author, Austin Connolly (1988)? Again, there is no item analysis and no factor analysis on this test.

Another tantalizing question is whether these content areas could be, in reality, strategy areas. In other words, are these authors representing as domains what are actually a variety of strategies or levels of competence? Therefore, what is measured could be a particular strategy adopted by a particular child, rather than an indication of the child's mathematics ability in a domain. For example, the "loading boxes" word problem described in the Lummis and Stephenson (1990) report requires visualization as well as verbal comprehension. Children may fail a test item either because they have poor oral comprehension skills, because they lack the ability to visualize the problem, or because they cannot represent the problem in mathematical terms. Furthermore, from the perspective of mathematics, what *is* the domain being represented? Is the "loading boxes" problem a particular domain of word problems as the author maintains, or is it a variant of a long division problem or of a visualization task? Which mathematics skills are critical?

The fundamental question of how many mathematics domains are truly distinct has rarely been asked, much less answered. There are various untested approaches to this question. Resnick (1991) presented one for the domain of arithmetic in which she isolated four subdomains: protoquantities, quantities, numbers, and operators and relations. Bruner (1973) set out a sequence for the evolution of expertise in mathematics, beginning with the prerule phase, to algorithms, to problem solving, and finally to invention. These levels of ability are intended to apply to each mathematics domain, but the domains remain unspecified.

Thus, the study of the psychology of mathematics is lacking in science's most basic requirement: the categorization of the domain of investigation. Instead, much of the research has focused on an analysis of how children and experts work at solving problems, rather than on a rigorous classification of the problems themselves. This could be a productive line of inquiry if the issues raised earlier are not neglected.

The groundwork for understanding the development of mathematics strategies was prepared by Piaget (1953) in his work on how children develop mathematical concepts. Unfortunately, after nearly 40 years, the impact of his insights on the pedagogy of mathematics has been minimal, and exemplary research on Piaget's basic ideas is still lacking.

Perhaps the most extensive work based on a Piagetian framework has been conducted by Davidson, a mathematician and educator. Her published work is represented largely by the adaptation, invention, and design of manipulative materials (Davidson, 1977; Davidson & Bennett, 1973; Davidson & Wilcutt, 1981, 1983). Davidson's applied work also provides a specific set of hypotheses that could form the basis of a research program.

The essence of Piaget's mathematics framework is that a child must begin with an understanding of the concrete properties of objects and with the fact that these can be ordered along specific dimensions. Instead of attaching a number system to the objects themselves, Piaget recommended that an intermediate step be introduced involving pictorial (two-dimensional) representations of the objects themselves. From this, the child learns that objects and properties of objects can be represented in an abstract form, but one that is still accessible. Numbers are introduced as the third step in the process. Davidson has brought in a fourth step in this sequence by moving from objects to counters and then to pictures of counters. For example, in her program called *Chip Trading* (Davidson, Galton, & Fair, 1975), she lets real chips stand for objects, then pictures of chips (colored dots) for the chips themselves, and finally, numbers to represent the pictures of colored dots.

The Chip Trading program has been developed to include alternative number systems (base 3, base 10) to create a real understanding of what a number system is, as well as special training in place value using the different bases.

Davidson is convinced that a child cannot really understand what a number system represents if the child is introduced only to numbers in isolation from any concrete or pictorial representation. Furthermore, she believes that merely using manipulatives by themselves without making a tangible connection to a number system is essentially fruitless. A child must be taught to move back and forth among numbers, pictures, and objects to grasp fully the principles involved.

Whereas Davidson and others using similar methods claim powerful results from this method, the actual research remains to be done. There are excellent reasons why research should follow her guidelines: (a) because we already know that teaching mathematics by numbers alone is ineffective; (b) because the method is based on Piaget's original research and has heuristic value; (c) because the materials are all available to do the research; and (d) because Davidson's own experience in teaching mathematics to children, teachers, and college students for more than 20 years has shown that these methods not only work, but generate enthusiasm for and interest in mathematics.

A second line of inquiry could be developed from Davidson's suggestion that there are two major learning styles in mathematical problem solving. To tie this in with the theme of this chapter, a research program that is based on Davidson's method of identifying the Type 1 and Type 2 learner would be expected, on the basis of the evidence reviewed earlier, to make strong predictions for sex differences in problem-solving strategies.

It is relevant to look at these problem-solving approaches in more detail. Follow the thinking of a Type 1 learner solving a problem in simple addition or multiplication. For the problem $8 + 6 = ?$, the child stores 8 away, and proceeds to count on by ones, that is, 9, 10, 11, 12, 13, 14, using his or her fingers to keep track. A more sophisticated version is to get to 10 (that takes 2), which leaves 4 fingers, and then $10 + 4 = 14$. Eventually, the solution is committed to rote memory: $8 + 6 = 14$. However, although the child may be able to move backward and forward in a counting system, he or she will not understand the fundamentals of a number system.

Another clue to the failure to comprehend a number system is to ask a child to count by tens. Most children can do this easily if they start from 10, but Type 1 children fall apart completely when asked to count by tens from some intermediate number like 17.

By contrast, a holistic Type 2 problem solver will reveal an understanding of a base 10 number system when solving problems. For instance, in the addition problem $8 + 6 = ?$, one solution is to collect the largest multiples of a base 10 system: $5 + 5$, with 3 and 1 left over, is $10 + 4 = 14$; or to collect terms using doubles, $6 + 6 = 12$, with 2 left over, $= 14$.

The contrast is more marked in a multiplication problem such as $9 \times 7 = ?$ The Type 1 child will interpret this as repeated addition: $7 + 7 + 7 + 7 + 7 + 7 + 7 + 7 + 7$. The child can usually get to 14, and then revert to fingers, or count forward while making slashes on a piece of paper. A Type 2 learner might use Cuisenaire rods or a centimeter measure, or discover the pattern of multiples in a hundreds chart (a matrix of 100 squares, numbered from 1 to 100). The pattern of multiples of 7 is to "go back three spaces $(10 - 3 = 7)$ and down one row."

Both types of learners will eventually reach the point at which these solutions are memorized and become automatic, but the Type 2 learner will have a much better grasp of what a number system represents and has the advantage of basing numbers on strong concrete and pictorial images.

So far, there is no empirical verification about these two types of learners. The tests are available to classify children by type, but the research has not been done, nor have the typologies outlined by Davidson been related to the teaching methods described earlier. Several interesting research questions stem from this line of thinking. Will the techniques outlined by Davidson or by Piaget turn Type 1 learners into Type 2 learners? Do these learning styles hold up over time? Do the teaching methods in the early grades make any difference to subsequent performance in more abstract mathematics? And pertinent to the issues raised in this chapter, will these methods eliminate any sex differences that may exist in learning style or in higher math performance?

Making Mathematics Interesting

Finally, what about the fact that females seem less interested in the world of inanimate objects? Does it matter? Problem solving can be made to be about something interesting and even about something living with a little ingenuity. Bransford (chapter 4) describes a program of research incorporating videos of real-life scenarios that have a series of mathematics problems imbedded in them. In one, an eagle will die from a gunshot wound if it cannot be rescued in time. To effect the rescue, the children have to determine which vehicle (jeep or plane) and which route to take. The distance that the plane can travel is determined by its weight plus that of the pilot and of the fuel. Its speed is determined by the velocity and direction of the wind. All clues are to be found somewhere in the video itself. Initial data indicate that both girls and boys become very involved with this problem-solving task and that the plight of the dying eagle must be especially motivating for girls.

In conclusion, we have the basis for radically different approaches to teaching mathematics skills, one that would create an enriched mental schema with which to represent numbers and mathematics concepts, and another that will make mathematics relevant and exciting to boys and girls alike. If both of these approaches were adopted in the classroom, sex differences in mathematics at all ages and all levels of ability would probably disappear. Meanwhile, research is needed from cognitive scientists to test the important contributions of people like Piaget and Davidson.

References

Barakat, M. K. (1951). A factorial study of mathematical abilities. *British Journal of Psychology, 4*, 137–156.

Barrett, E. S. (1955). The space visualization factors related to temperamental traits. *Journal of Psychology, 39*, 279–287.

Benbow, C. P. (1988). Sex differences in mathematical reasoning ability in intellectually talented preadolescents: Their nature, effects, and possible causes. *Behavior and Brain Sciences, 11*, 169–232.

Benbow, C. P., & Stanley, J. C. (1983). Sex differences in mathematical reasoning ability: More facts. *Science, 22*, 1029–1031.

Bradley, L., & Bryant, P. E. (1983). Categorizing sounds and learning to read: A causal connection. *Nature, 301*, 419–421.

Bronowski, J. (1973). *Ascent of man.* Boston: Little, Brown.

Bruner, J. S. (1973). *Beyond the information given.* New York: Norton.

Brush, S. G. (1991). Women in science and engineering. *American Scientist, 79*, 404–419.

College Entrance Examination Board. (1987). *College bound seniors: Profile of SAT and achievement test takers.* New York: Author.

Connolly, A. J. (1988). *Key Math-R.* Circle Pines, MN: American Guidance Service.

Davidson, P. (1977). *Idea book for the use of cuisenaire rods at the primary level.* White Plains, NY: Cuisenaire Company of America.

Davidson, P., & Bennett, A. B. (1973). *Fraction bars program.* Ft. Collins, CO: Scott Resources.

Davidson, P., Galton, G. K., & Fair, A. W. (1975). *Chip trading activities.* Ft. Collins, CO: Scott Resources.

Davidson, P., & Wilcutt, R. E. (1981). *From here to there with Cuisenaire rods: Area, perimeter and volume.* White Plains, NY: Cuisenaire Company of America.

Davidson, P., & Wilcutt, R. E. (1983). *Spatial problem solving with cuisenaire rods.* White Plains, NY: Cuisenaire Company of America.

Fennema, E., & Sherman, J. (1977). Sex-related differences in mathematics achievement, spatial visualization and affective factors. *American Educational Research Journal, 14*, 51–71.

Feshbach, N. D., & Hoffman, M. A. (1978, April). *Sex differences in children's reports of emotion arousing situations.* Paper presented at the meeting of the Western Psychological Association, San Francisco.

Goswami U., & Bryant, P. (1990). *Phonological skills and learning to read.* Hillsdale, NJ: Erlbaum.

Goodenough, E. W. (1957). Interest in persons as an aspect of sex differences in the early years. *Genetic Psychological Monographs, 55*, 287–323.

Harris, L. J. (1978). Sex differences in spatial ability: Possible environmental, genetic and neurological factors. In M. Kinsbourne (Ed.), *Hemispheric asymmetries of function.* Cambridge, England: Cambridge University Press.

Harvey, J., & Goldstein, S. (1985). Sex differences in science and mathematics for more able pupils. *Gifted Education International, 3*, 133–136.

Hyde, J. S., Fennema, E., & Lamon, S. J. (1990). Gender differences in mathematics performance: A meta-analysis. *Psychological Bulletin, 107*, 139–155.

Jahoda, G. (1979). On the nature of difficulties in spatial–perceptual tasks: Ethnic and sex differences. *British Journal of Psychology, 70*, 351–363.

Liben, L. S., & Golbeck, S. L. (1984). Performance on Piagetian horizontality and verticality tasks: Sex-related differences in knowledge of relevant physical phenomena. *Developmental Psychology, 30*, 595–606.

Liberman, I. Y., & Mann, V. (1981). Should reading instruction and remediation vary with the sex of the child? In A. Ansara (Ed.), *Sex differences in dyslexia* (pp. 151–167). Towson, MD: Orton Dyslexia Society.

Lummis, M., & Stevenson, H. W. (1990). Gender differences in beliefs and achievement: A cross-cultural study. *Developmental Psychology, 26*, 254–263.

Maccoby, E. E., & Jacklin, C. N. (1974). *The psychology of sex differences*. Stanford, CA: Stanford University Press.

McGuinness, D. (1985). *When children don't learn*. New York: Basic Books.

McGuinness, D. (1990). Behavioral tempo in pre-school boys and girls. *Journal of Learning and Individual Differences, 2*, 315–326.

McGuinness, D., & McLaughlin, L. (1982). An investigation of sex differences in visual recognition and recall. *Journal of Mental Imagery, 6*, 203–212.

McGuinness, D., & Morley, C. (1991). Sex-differences in the development of visuo-spatial ability in preschool children. *Journal of Mental Imagery, 15*, 143–150.

McGuinness, D., Olson, A., & Chaplin, J. (1990). Sex differences in incidental recall. *Journal of Learning and Individual Differences, 2*, 263–286.

McGuinness, D., & Sparks, J. (1983). Cognitive style and cognitive maps: Representations of a familiar terrain. *Journal of Mental Imagery, 7*, 91–100.

McGuinness, D., & Symonds, J. (1977). Sex differences in choice behavior: The object–person dimension. *Perception, 6*, 691–694.

O'Halloran, P. J. (1983). Sex differences in mathematics. *New Zealand Journal of Educational Studies, 18*, 188.

Pattison, P., & Grieve, N. (1984). Do spatial skills contribute to sex differences in different types of mathematical problems? *Journal of Educational Psychology, 76*, 678–689.

Piaget, J. (1953). How children form mathematical concepts. *Scientific American, 189*, 74–79.

Rayner, K., & Pollatsek, A. (1989). *The psychology of reading*. Englewood Cliffs, NJ: Prentice Hall.

Resnick, L. (1991, November). Presentation to Contributions of Psychology to Mathematics and Science Education, Tampa, FL.

Stallings, J. A. (1979, September). *Comparison of men's and women's behavior in high school math classes*. Paper presented at the 87th Annual Convention of the American Psychological Association, New York. (Also reprinted as a bulletin of the Stanford Research Institute, 1979)

Thurstone, L. L. (1950). *Some primary abilities in visual thinking*. Chicago: University of Chicago Press.

Trabasso T., & Bower, G. H. (1968). *Attention in learning: Theory and research*. New York: Wiley.

Wagner, R. K., & Torgesen, J. K. (1987). The nature of phonological processing and its causal role in the acquisition of reading skills. *Psychological Bulletin, 101*, 192–212.

Zeidner, M. (1986). Sex differences in scholastic aptitude: The Israeli scene. *Personality and Individual Differences, 7*, 847–852.

PART FOUR

Applied Psychological Research on Mathematics and Science Education

Applied Psychological Research on Mathematics and Science Education

Howard M. Knoff

When considering the application of psychological research in the areas of mathematics and science education to the classroom, two important perspectives are critical: the scientist–practitioner perspective, and the perspective that puts this research into an empirical context. The scientist–practitioner perspective is needed because much of this research has been devoted to demonstrating the validity of specific research hypotheses, whereas investigations focusing on the application of this research in actual classroom settings as a long-standing and fundamental part of the classroom routine or as the teacher's mode of instruction have not been as prevalent. Expanding briefly, scientists–practitioners must recognize and attend to the following principles to maximize their use of research in this area:

1. They should understand the theoretical and empirical base underlying mathematics and science research.

2. They should understand the goal of the research (e.g., to extend, validate, or replicate a theoretical or empirical principle; to explore the

impact of a population, a sample characteristic, or some other intervening variable; to evaluate the impact of a specific intervention or research application; or to assess the conditions or variables needed to implement an intervention design efficaciously).

3. They should be critical consumers of the research, its methodology, its statistical analyses, its results, and the implications drawn from these results.

4. They should evaluate independently the generalizability of the research to different populations, samples, or individual students. Here, the evaluation of generalizability is especially important when attempting to apply specific interventions, tested experimentally with large groups, to individual students in real classroom settings.

5. They should evaluate independently the generalizability of the research to settings with fewer environmental controls and resources, and they should consider the steps to be taken to ensure treatment integrity.

Attention to these principles facilitates the necessary discrimination between research that is ongoing, programmatic, and not ready for large-scale implementation, and research that is ready to be tested (or fully implemented) in real settings under actual conditions. Without this attention, research findings might be applied inappropriately or prematurely with individual students in a classroom setting.

Even when carefully attending to these principles, the scientist–practitioner needs to put the application of this research into a pragmatic, yet empirical, context. This context has been provided by Centra and Potter (1980), Rosenfield (1987), McKee and Witt (1990), and Knoff and Batsche (1991), who have developed an empirical model focused on the four student learning outcomes typically desired in school settings: academic skill outcomes, student cognitive/metacognitive skill outcomes, social skill outcomes, and adaptive behavior skill outcomes. This model suggests that these four skill outcomes occur because of the interdependent interaction of home, school, and student variables and conditions.

The home variables actually involve family, neighborhood, and community conditions that can be defined in the following way:

Family, Neighborhood, and Community Conditions. These involve characteristics and conditions of a referred student's family, neighborhood, and community as they relate ultimately to effective teaching and to student learning outcomes. They emphasize the importance of a healthy home environment and its impact on a student's school readiness and success (Knoff & Batsche, 1991, pp. 177, 180).

The school variables concern school and school district conditions, within-school and specific classroom conditions, teacher/instructional conditions, and curricular conditions. These can be defined in the following way:

School/School District Conditions. These involve conditions such as the physical plant, presence of resources that support learning, and programs in the school that reinforce professional development and teaching excellence.

Within School/Classroom Conditions. These involve conditions of a student's school building and actual classroom, such as the administrative organization of the building and the instructional organization of the school and classroom, as well as within-classroom conditions that explain teacher effectiveness.

Teacher Instructional/Conditions. These involve characteristics and conditions that teachers bring to the classroom and that ultimately translate into effective instructional skills and behaviors affecting student learning (e.g., background characteristics, professional training). These also include those empirically identified skills, activities, and conditions that teachers perform to make their instruction effective and meaningful (e.g., their individualization of the curriculum for specific students, their ability to adapt instruction to diverse learning styles and circumstances).

Curricular Conditions. These involve characteristics and conditions of the curricula being used and include content of the curriculum as well as the pedagogical processes that it uses to ensure student learning and mastery (Knoff & Batsche, 1992, p. 180).

The student variables encompass those characteristics and conditions that relate to an individual's readiness for learning and those that are actually used during the learning process. These two areas can be defined in the following way:

Students' Readiness Conditions. These involve often preexisting characteristics and conditions that relate primarily to a student's cognitive and academic ability, educational attitudes, and readiness for academic and social learning. These characteristics relate directly to those academic behaviors that support learning progress and achievement.

Students' Applied Learning Conditions. These involve the behaviors that students exhibit and that directly support their academic and social learning and progress, such as self-competence skills, social skills, and effective use of teachers and of teacher time (e.g., academic engagement; Knoff & Batsche, 1991, pp. 180–181).

The integration of these conditions into a single model is represented in Figure 1.

The importance of this empirical context rests with the scientist–practitioner who must evaluate existing mathematics and science research and decide where it fits in the comprehensive model described earlier. For example, does the research address only one facet of the model (i.e., individual home, school, or student variables), does it address an interaction of two conditions (e.g., home–school, home–student, or school–student interactions), or does the research address a complex, interactive effect that transcends all three primary areas described? Furthermore, does the research address situations that affect most students in normative and developmental ways, or is it focused on an area that taps into more idiographic characteristics affecting students' individual learning outcomes? Finally, what is the generalizability of the research, and can it be implemented with treatment integrity?

This section of the book highlights three experts' perspectives on how psychological research in mathematics and science can be applied to education.

McCombs first discusses a set of learner-centered principles, developed by the American Psychological Association's Task Force on Psychology in Education in 1991 as its contribution to the school reform movement that swept the country with the *America 2000* initiative. These principles take into account long-standing research findings that address those psychological factors that are primarily internal to the learner and yet affect the entire learning process and classroom environment. Although these principles clearly focus on students' readiness and on the student applied learning conditions discussed earlier, they fully acknowl-

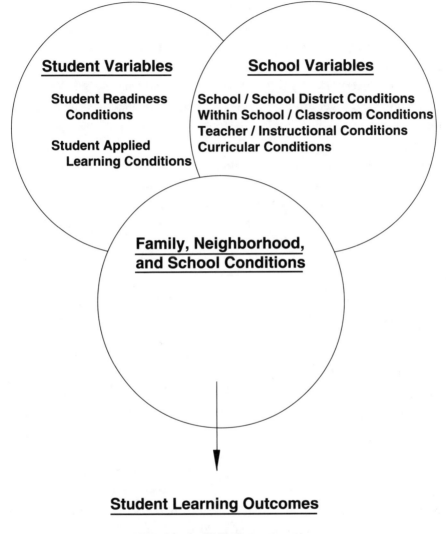

FIGURE 1. Empirically based model of variables and conditions that lead to student learning outcomes.

edge that these conditions interact with the external environment and with other contextual factors. A major contribution in the "principles" document is the fact that each principle is operationalized to address learning and instructional implications.

After describing the 12 principles (organized according to metacognitive and cognitive, affective, developmental, social, and individual-differences research), McCombs discusses their implications for mathematics and science education. Initially, she focuses on the learner-centered perspective that she defines as "one that reciprocally focuses on promoting the motivation and higher level thinking of students and the adults who interact with them." Once again, the interdependent interaction of home, school, and student variables and conditions is emphasized, as is the importance of using instructional approaches and curricula that take advantage of children's natural curiosity and of their desire to learn and explore.

From an "evaluation of learning" perspective, McCombs maintains her learner-centered focus by advocating for cooperative and collaborative learning activities, tasks that teach and facilitate student control and self-regulation, and assessment approaches that are authentic. McCombs cites a presentation by Marzano and Kendall, who feel that learning should be data-based; multidimensional; interdisciplinary; interactional; student-directed; production-oriented; and personally, socially, and domain relevant. She then defines the outcomes of learning as involving students' ability to think, self-regulate, work cooperatively, and master specific content areas. McCombs closes her chapter with recommendations for three new research and development directions, each of which should make it easier for learners who are "first in the world in mathematics and science achievement by the year 2000."

McCombs' chapter provides an excellent overview and orientation to this section of the book. By understanding the psychological research underlying the 12 learner-centered principles, one can understand the critical impact children have on their own learning. Understanding the learning and instructional implications of each principle makes it possible to begin to develop curricular systems, at home and at school, that can foster integrated learning environments, thus making the learning process successful and life long.

In the next chapter, Daniel Keating discusses the developmental diversity of children as it relates to mathematical and scientific competence. In the first part of his chapter, he explains that he uses the term *developmental diversity* as an alternative to the term *individual differences* because (a) it is more neutral and can encompass the biological and socialization effects of child development and (b) it suggests a more multivariate description of a population, rather than being a focused search for pathology. Then, he conceptually defends his belief that an understanding of the diverse developmental histories that explain why different individuals and groups attain competence can foster a broader and more generalized understanding of the critical features that contribute to the mathematics and science competence of all children. Finally, he provides an overview of the key ideas from the psychometric, Piagetian, information-processing, and Vygotskian perspectives, focusing away from the deficit- and product-oriented position of the psychometric models and toward the developmental, process-oriented, and contextualistic positions of the latter three perspectives.

In using the four perspectives noted earlier toward an integrated understanding of the development of competence, Keating talks about the need to investigate cognitive activities (e.g., motivation, effort, aspiration, emotion), instead of mental abilities, as the basic building blocks. However, he emphasizes that these cognitive activities must be understood in the context of social and developmental processes (Keating uses the term *cognitive socialization*), and of their interaction and occurrence within the instructional environment. To do this, Keating suggests that numerous developmental pathways exist and should be investigated and that "because variability typically has been explored within a small number of more normative pathways to the development of expertise, diversity or difference has often been equated with deficit." Keating eschews a univariate view and advocates a multivariate exploration, including an exploration of competence as demonstrated in Asian and European countries and cultures.

Finally, borrowing from the biological developmental study of anomalies, Keating discusses the usefulness of researching why some students are not competent in mathematics and science. Here, he recommends investigations of such developmental skills as speed of processing, declarative knowledge, procedural knowledge, logical understanding, and cognitive self-regulation, and notes that students' integration of conceptual knowledge is not automatic, concluding that an Instruction × Child interaction might hold the best explanation. Keating closes his chapter by stating that four features (content knowledge, critical thinking, creative thinking, and communication) may be viewed as the keys to mathematical and scientific competence. This perspective, however, is suggested with a caveat that the timing of instruction and intervention may be at least as important as the content of these curricular areas.

Keating does an admirable job of addressing the child–instruction (i.e., teacher)–curriculum interaction that translates into mastery learning (and competence) for most children. Although he does not discuss substantially the impact of home, neighborhood, and community variables, his ecological approach and cognitive socialization perspective would appear to include them as necessary. Overall, Keating delivers a very strong message that contradicts much of the educational assessment mindset of the late 20th century: We should stop categorizing students' curricular behavior along the so-called "normal" curve, and begin to look at individual student's developmental progress in mathematics and science and at the cognitive and instructional processes that maximize that progress.

In the final chapter of this section, Robert Siegler writes specifically about the adaptive and nonadaptive strategies that low-income children use to solve elementary-school-level mathematics problems. Noting that most of the research in this area has investigated middle-income children, Siegler first presents four generalizations with these individuals: (a) Elementary school students use diverse strategies, both individually and at a given age, to solve different types of mathematics problems. (b) These students choose adaptively among these diverse strategies, often using "backup" strategies when problems are more difficult and accuracy is needed. (c) Strategy choice varies because of students' individual differences. (d) A student's ability to execute a backup strategy and his or her confidence with retrieving an answer best predict the actual retrieval of a calculation problem's correct answer. Siegler then describes a series of research studies with low-income elementary students.

In summarizing the results of these studies, Siegler notes that (a) children from low-income families also choose their mathematics strategies adaptively; (b) they exhibit the same types of individual differences as children from middle-class backgrounds; (c) they follow the same predictive model for calculation retrieval as all other children; and (d) an understanding of these students' strategy use, strategy choice, and strategy comprehension holds the key to the most effective instruction. These results indicate that low-income students are very similar to middle-income students in the way in which they solve mathematics problems and that the issue is more one of teaching and exposure than of ability and motivation. Finally, Siegler believes strongly that children in need of mathematics instruction need help executing backup strategies more effectively and that their success in using backup strategies effectively leads to later immediate retrieval of calculation answers without the need for backup strategies.

Although Siegler's chapter focuses a great deal on school readiness and school-applied learning conditions, it has direct implications for both instruction and the adaptation and effective use of mathematics curricula. Perhaps more significantly, however, Siegler's research helps to respond to the "excuse" often heard in schools that the (low-income) family environment is responsible for a specific child's inability to learn. Clearly, with all children, instruction can be the great moderator of unfortunate home circumstances. However, with low-income students, instruction often is the only moderator—one that, if unsuccessful, results in school failure, school dropout, and the loss of an individual to begin the process anew with the next generation.

The chapters in this section of the book potentially have the greatest impact on attaining the *America 2000* goals. If we are unable to apply psychology to mathematics and science education and to other significant aspects of the schooling process, our children will never (a) start school ready to learn, (b) graduate from high school with competence and functional skill in the basic academic areas, or (c) be ready to succeed in higher education or in the skilled workforce. McCombs, Keating, and Siegler have provided a diverse, yet detailed, road map to the future. We must all consult this road map, and others, to begin the process of psychological application toward a destination of educational excellence.

References

Centra, J. A., & Potter, D. A. (1980). School and teacher effects: An interrelational model. *Review of Educational Research, 50,* 273–291.

Knoff, H. M., & Batsche, G. M. (1991). Integrating school and educational psychology to meet the educational and mental health needs of all children. *Educational Psychologist, 26,* 167–183.

McKee, W. T., & Witt, J. C. (1990). Effective teaching: A review of instructional and environmental variables. In T. Gutkin & C. Reynolds (Eds.), *The handbook of school psychology* (2nd ed., pp. 821–846). New York: Wiley.

Rosenfield, S. (1987). *Instructional consultation.* Hillsdale, NJ: Erlbaum.

CHAPTER 9

Learner-Centered Psychological Principles for Enhancing Education: Applications in School Settings

Barbara L. McCombs

The era of radical educational reform is upon us. No longer are educators and psychologists aiming to improve the existing system. Rather, there is widespread agreement that the existing system is not working and that basic assumptions underlying this system are fundamentally flawed (e.g., Covington, 1991; Eisner, 1991; Heshusius, 1991; Hutchins, 1990; Levin, 1991; Skrtic, 1991). As a result, a flurry of activity is taking place at national, state, and local levels to address the redesign of curriculum, instruction, and the entire enterprise of schooling as we know it. A large number of these efforts acknowledge, at least implicitly, that comprehensive, learner-centered, and systemic redesign strategies are required to address holistically the unique needs of individuals, the family, and the social systems of which they are a part, and the educational and administrative systems that interface with the learner and with his or her supporting social systems. In addition, strategies for improving student achievement are being viewed within the larger context of transformed ideas on teaching and learning.

The past century of psychological research has contributed significantly to these transformed views on teaching and learning. From the areas of human development, learning, cognition, and motivation, research in psychology relevant to education has begun to be integrated in ways that can contribute more directly not only to improved practice, but also to new models or paradigms of education. A set of learner-centered psychological principles for guiding school redesign and reform was drafted by the American Psychological Association's Task Force on Psychology in Education (McCombs, 1992). These principles represent psychology's accumulated knowledge base with implications for learning and instruction within and across content areas represented in school curricula. They also are based on the belief that improvements in educational practice will occur only when the system is redesigned with primary focus on the learner. Together, these principles represent a comprehensive integrative framework that can lead to alternative, learner-centered models of schooling to challenge teachers to understand students of different backgrounds and abilities, their concepts of what it means to perform at their best, and their values and standards regarding educational achievement.

Examples of learner-centered redesign efforts at all levels of the system are beginning to surface. The alternative-school literature provides one set of examples of successful learner-centered models for the entire process of schooling for students most at risk of school failure (e.g., Loria, 1989; Smith, Gregory, & Pugh, 1981). At the curriculum level, the past 10 years have seen a number of efforts to develop curricula and instructional practices to enhance students' higher level thinking and reasoning skills (e.g., Callahan, Pogrow, & Gore, 1990; Gore & Pogrow, 1991; Marzano et al., 1988). In general, these programs either teach higher level thinking skills as a separate curriculum or they embed strategies for higher level thinking and problem solving within the context of specific content areas such as mathematics and science. Program results have been positive, particularly for students considered at risk educationally (Callahan et al., 1990).

In spite of examples of significant progress, however, there has been surprisingly little consensus about educational goals, competencies, and standards, or about how these should be accomplished and assessed in ways that respect individual differences and help diverse groups of students realize their potentials. Even though there is clear evidence that learner-centered approaches can promote maximum student involvement and learning outcomes, not all educators and policymakers are convinced that this is the direction in which to go to improve educational outcomes. Many still argue for changes at the instructional or administrative level of the educational system (e.g., tougher standards, more classroom controls, standardized curricula). These suggestions, however well-intended, are based on assumptions regarding the learner and the learning process that overlook his or her basic nature and needs in learning activities. Thus, the realization of our shared educational goals requires strategies for transforming traditional assumptions and perspectives to arrive at a new paradigm for educational systems.

The purpose of this chapter is to present a set of learner-centered psychological principles that can guide curriculum and instructional reforms in mathematics and science education. Specific implications of these principles are discussed, followed by examples of how the principles have been integrated within innovative intervention programs implemented in school settings. In this context, I give particular attention to my work and that of my colleagues at the Mid-continent Regional Educational Laboratory. The presentation concludes with implications for further school and classroom research and development that can enhance students' motivation and achievement in mathematics and science.

Learner-Centered Principles for Curriculum and Instruction

Among those questioning fundamental assumptions underlying current practice in curriculum and instruction, Heshusius (1991) examined critically the ontological and epistemological beliefs within these assumptions. She argued convincingly that they are based on a mechanistic and reductionistic worldview that does not fit within contemporary views of teaching and learning. To reflect current views of the holistic, self-regulatory, purposeful, constructive, and intrinsically motivated nature of learning requires a learner-centered paradigm. Heshusius (1991) stated,

> to be child directed, any set of constructs and procedures for instruction and assessment should directly emerge ... from an understanding of how children perceive and construct their world on their terms, and from knowledge of how social and cultural processes shape the very ways a person constructs knowledge. (p. 320)

In this postpositivist era, we are redefining what it means to be scientific about human behavior. In Heshusius's view, fact cannot be separated from value, the knower from that which is known, or cognition from meaning and affect. The science of human behavior cannot take the person (with his or her values, interests, needs, and consciousness) out of the act of knowing (Heshusius, 1991). Furthermore, *learner-* versus *learning*-centered approaches are fundamentally different. Learner-centered approaches take the learner's perspective into account in the design of educational systems. Learning-centered approaches often redesign the system for the learner from the perspective of the instructional designer.

The numbers of the APA Task Force on Psychology in Education could not agree more. Very similar concerns prompted the task force to derive a set of learner-centered psychological principles that could provide guidelines for school redesign and reform. These guidelines are seen as a way to lay out an entire new set of assumptions underlying a learner-centered paradigm for curriculum and instruction. In addition, the principles are not only consistent with more than a century of psychological research, but they are also widely shared and implicitly recognized in many excellent programs found in today's schools. They integrate the combined experience of research and practice in a variety of psychological areas including educational, school, clinical, developmental, experimental, social, organizational, community, and engineering psychology. In addition, the principles reflect an integration of both conventional and scientific wisdom that can lead to effective schooling, as well as to the positive mental health and to more effective functioning of our nation's children, their teachers, and the organizational systems that serve them.

As developed by the APA Task Force, the learner-centered principles take into account psychological factors that are primarily internal to the learner while recognizing external environmental or other contextual factors that interact with these internal factors (McCombs, 1992). The first 10 principles subdivide into those referring to metacognitive and cognitive, affective, developmental, and social factors. Two final principles cut across the prior principles and focus on what we know about individual differences. The entire list of 12 principles is provided, along with some of the learning and instructional implications for each principle that are relevant to mathematics and science education.

Metacognitive and Cognitive Factors

Principle 1
Learning is naturally an active, volitional, internally mediated, and individual process of constructing meaning from information and experience, filtered through each individual's unique perceptions, thoughts, and feelings.

Learning Implications

- Students have a natural inclination to learn and will assume personal responsibility for learning (e.g., monitoring; checking for understanding; and becoming an active, self-directed learner) in an environment that takes past learning into account, links new learning to personal needs, and actively engages students in their own learning process.

Instructional Implications

- Effective instruction focuses on the active involvement of students in their own learning, with opportunities for teacher and peer interactions that engage students' natural curiosity.
- Effective instruction encourages students to make meaningful links between prior knowledge and new information by providing multiple ways of presenting and representing information (e.g., auditory, visual, kinesthetic).

Principle 2
The learner seeks to create internally consistent, meaningful, and sensible representations of knowledge regardless of the quantity and quality of data available.

Learning Implications

- Learners generate integrated, "commonsense" representations and explanations even for poorly understood or communicated facts, concepts, principles, or theories.

Instructional Implications

- Effective instructional tasks and materials engage the "whole learner" and incorporate learning assessments that can be used by both students and teachers to check for student understanding of the subject matter.
- Effective instructional practices include diagnosis of students' knowledge and understanding in various content areas, and use of diagnostic results to inform teaching and the selection of appropriate instructional materials.

Principle 3
The learner organizes information in ways that associate and link new information with existing knowledge in uniquely meaningful ways.

Learning Implications

- Learning can be facilitated by assisting learners in acquiring and integrating knowledge (e.g., by teaching strategies for constructing meaning, relating new knowledge to general themes or principles, accessing prior knowledge, organizing content, and storing or practicing what they have learned).

Instructional Implications

- Effective instruction attends both to the content of curriculum domains and to generalized and domain-specific process strategies for acquiring and integrating knowledge in these domains, including opportunities for practicing a variety of learning strategies in different content domains.

- Effective instruction includes a concern with constructive and informative feedback regarding the learner's instructional approach and product, as well as sufficient opportunities to practice and apply knew knowledge and skills to appropriate levels of mastery.

Principle 4

Higher order strategies for "thinking about thinking" (e.g., for overseeing and monitoring mental operations) facilitate creative and critical thinking and the development of expertise.

Learning Implications

- Learners are capable of a metacognitive or executive control level of thinking about their own thinking that includes self-awareness, self-monitoring, and self-regulation of the processes and contents of thoughts, knowledge structures, and memories. Self-awareness of personal agency promotes commitment and persistent implementation of learners' intentions and goals.

Instructional Implications

- Effective instructional materials and curricula provide explicit opportunities for students to engage metacognitive capacities and practice metacognitive strategies, including reflective self-awareness and goal setting.
- Effective instructional practices encourage problem solving, debates, group discussions, and other strategies that enhance the development of higher order thinking and use of metacognitive strategies.

Affective Factors

Principle 5

The depth and breadth of information processed, and what and how much is learned and remembered, are influenced by (a) self-awareness and beliefs about one's learning ability (personal control, competence, and ability); (b) clarity and salience of personal needs; (c) personal expectations for success or failure; (d) affect, emotion, and general states of mind; and (e) the resulting motivation to learn.

Learning Implications

- Learners have a rich internal context of beliefs, goals, expectations, feelings, and motivations that can enhance or interfere with the quality of thinking and information processing. The relationships among thoughts, mood, and behavior underlie students' psychological health and functioning as well as their learning efficacy. Factors such as low reflective self-awareness; negative personal beliefs; lack of personal learning goals; negative expectations for success; and anxiety, insecurity, or pressure that make learning aversive interfere with optimal learning.

Instructional Implications

- Effective learning materials, activities, and experiences have an affective and cognitive richness to help students generate positive thoughts and feelings of excitement, interest, and stimulation.
- Effective curricula attend to affect and to mood, as well as to cognition and thinking in all learning activities and experiences.

Principle 6
In the absence of intense negative cognitions and emotions (e.g., insecurity, worrying about failure, being self-conscious), individuals are naturally curious and enjoy learning.

Learning Implications

- Positive motivation for learning is dependent largely on helping to bring out and develop students' natural curiosity or intrinsic motivation to learn, rather than on "fixing them" or giving them something that they lack.

Instructional Implications

- Effective instructional practices maintain fair, consistent, and caring policies that respect individual students and maintain a safe climate for learning—one that focuses on individual mastery versus competitive performance goals.
- Effective instructional materials and practices include strategies that elicit or stimulate students' intrinsic motivation to learn and

that avoid an overreliance on external rewards that can undermine natural learning interest.

Principle 7
Curiosity, creativity, and higher order thinking processes are stimulated by learning tasks of optimal difficulty, relevance, authenticity, challenge, and novelty for each student.

Learning Implications

- Effective curricula include "authentic" tasks and assessments that help students to integrate information and performance across subject matter disciplines while allowing students to choose appropriate levels of difficulty, challenge, novelty, and so forth.
- Effective instructional practices encourage student choice in areas such as topics of learning, types of projects to work on, or whether to learn independently or in groups.

Developmental Factors

Principle 8
Individuals proceed through discrete, identifiable progressions of physical, intellectual, emotional, and social development that are a function of unique genetic and environmental factors.

Learning Implications

- Children learn best when the material is appropriate to their developmental level and presented in an enjoyable, interesting way while at the same time challenging their intellectual, emotional, physical, and social development.

Instructional Implications

- Effective instructional materials and curricula are developmentally appropriate to the unique intellectual, emotional, physical, and social characteristics of students.
- Effective instructional practices are flexible in matching individual student needs with variations in instructional format and processes, including content, structure, strategies, and social settings.

Social Factors

Principle 9
Learning is facilitated by social interactions and communication with others in a variety of flexible, diverse (cross-age, culture, family background, etc.), and adaptive instructional settings.

Learning Implications

- Learning is facilitated by including diverse settings that allow the learner to interact with a variety of students from different cultural and family backgrounds, interests, and values. Divergent and flexible thinking, as well as social competence and moral development, are encouraged in learning settings that allow for and respect diversity.

Instructional Implications

- Effective strategies for grouping students for learning activities provide for an appropriate diversity of abilities, ages, cultures, and other stable individual differences.
- Effective instructional practices include those that attend to meaningful performance contexts (e.g., apprenticeship settings) wherein knowledge can be contextualized and anchored to meaningful and relevant prior knowledge and experience.

Principle 10
Self-esteem facilitates learning and is heightened when individuals are in respectful and caring relationships with others who see their potential, genuinely appreciate their unique talents, and accept them as individuals.

Learning Implications

- Individuals' access to higher order, healthier levels of thinking, feelings, and behaving is facilitated by quality personal relationships. Teachers' states of mind, stability, trust, and caring are preconditions for establishing a sense of belonging and a positive climate for learning.

Instructional Implications

- Effective instructional practices are those that foster quality adult–student and student–student relationships. Relationships that are based on understanding and mutual respect reciprocally reduce levels of stress and insecurity in teachers and students.

- Effective learning environments are warm, comfortable, and supportive. They provide a climate in which students' insecurities are alleviated and in which they feel a sense of belonging.

Individual Differences

Principle 11

Although basic principles of learning, motivation, and effective instruction apply to all learners regardless of ethnicity, race, gender, presence or absence of physical handicap, religion, and socioeconomic status, learners differ in their preferences for learning mode and strategies, the pace at which they learn, and unique capabilities in particular areas.

Learning Implications

- The same basic principles of learning, motivation, and effective instruction apply to all learners. At the same time, however, learners have unique capabilities and talents, and have acquired different preferences for how they like to learn and for the pace at which they learn. In addition, it must be recognized that learning outcomes are an interactional and interdependent function of student differences, as well as curricular and environmental conditions.

Instructional Implications

- Effective instructional practices ensure that all students have experience with (a) teachers interested in their area of instruction, (b) positive role modeling and mentoring, (c) constructive and regular student evaluations, (d) high teacher expectations, and (e) use of questioning skills to involve them actively in the learning process.
- Effective learning environments provide equal standards and requirements for all students while also showing respect for cultural diversity and developmental and other individual differences.

Principle 12

Beliefs and thoughts, resulting from prior learning and based on unique interpretations of external experiences and messages, become each individual's basis for constructing reality or interpreting life experiences.

Learning Implications

- Unique cognitive constructions form a basis for beliefs about and attitudes toward others. Awareness and understanding of these phenomena allow greater choice in what one believes, more control over the degree to which one's beliefs influence one's actions, and an ability to see and take into account others' points of view. The cognitive and social development of a child and the way in which a child interprets life experiences is a product of prior schooling and of home and community factors.

Instructional Implications

- Effective curriculum materials and activities help students to increase self-awareness and understanding and to understand different individuals, as well as different social and religious groups.
- Effective curricula also include activities that promote empathy and understanding, and respect for individual differences and for different perspectives.

Taken together, these principles and their respective implications for curriculum and instruction have specific implications for mathematics and science education, which are explored in the next section.

Implications for Mathematics and Science Education

Mathematics and science education have become a particular target for educational reform in our nation, as evidenced by a plethora of reports and by national and state goals. In fact, former President Bush and the nation's governors set the goal that American students will be first in the world in mathematics and science achievement by the year 2000. For that to happen, it is critical that learner-centered principles define the primary paradigm for mathematics and science education. What does that mean? On the basis of the preceding 12 principles, I believe it means that to maximize student learning, motivation, and achievement in these fields, attention needs to be given to at least the following dimensions in curriculum and instruction reforms:

- active involvement of students in their own learning;
- multiple modes of information presentation;

- frequent opportunities for teacher and peer interaction;
- diagnosis and selection of instructional materials and methods based on each student's unique knowledge representation;
- ongoing assessments of learning by and for teachers and students;
- individual and group opportunities for the development of higher level thinking, problem solving, goal setting, and reflective self-awareness;
- integration of affective strategies for stimulating interest and curiosity, and for enhancing perceptions of competence and motivation to learn;
- focus on individual learning and mastery versus competitive performance goals;
- minimal use of external rewards and incentives;
- use of "authentic" content topics perceived by students to be personally and socially relevant;
- developmentally appropriate opportunities for student choice and self-regulation;
- heterogeneous and diverse groupings of students for learning activities;
- meaningful performance contexts comparable to real-world settings;
- quality interpersonal relationships based on understanding and on mutual respect between students and teachers, and students and students;
- equal standards and requirements for all students while also respecting cultural diversity and other individual differences.

The foregoing considerations can become a "yardstick" against which to evaluate the degree to which various curricula and practices are learner centered. Such an evaluation also needs to take into account and to be integrated with new content paradigms being set forth by the mathematics and science professions. For example, the National Council of Teachers of Mathematics (NCTM) standards, the Mathematical Sciences Education Board (MSEB), and the American Association for the

Advancement of Science (AAAS) have set out the following new standards and content frameworks:

> Knowing mathematics means being able to use it in purposeful ways. To learn mathematics, students must be engaged in exploring, conjecturing, and thinking rather than only in rote learning of rules and procedures. Students should learn to value mathematics, reason mathematically, communicate mathematically, become confident of their mathematical abilities, and become mathematical problem solvers. (NCTM, 1989)

> The focus of school mathematics is shifting from a dualistic mission—minimal mathematics for the majority, advanced mathematics for a few—to a singular focus on a significant common core of mathematics for all students. The teaching of mathematics is shifting from an authoritarian model based on "transmission of knowledge" to a student-centered practice featuring "stimulation of learning." The teaching of mathematics is shifting from preoccupation with inculcating routine skills to developing broad-based mathematical power. The teaching of mathematics is shifting from an emphasis on tools to future courses to greater emphasis on topics that are relevant to students' present and future needs. (MSEB, 1990, p. 5)

> The basic dimensions of scientific literacy are being familiar with the natural world and recognizing both its diversity and unity, understanding key concepts and principles of science, being aware of some of the important ways in which science, mathematics and technology depend upon one another, knowing that science, mathematics, and technology are human enterprises and know what that implies about their strengths and weaknesses, having a capacity of scientific ways of thinking for individual and social purposes. (AAAS, 1989, p. 4)

As is being recognized to at least some degree by curriculum reformers, these new content frameworks need to be integrated with psychological principles regarding how and why learners learn and with the conditions that promote optimal learning and motivation. Both the nature of the task in which content is embedded and the nature of interactions between students and teachers and among students are essential to achieving improved student outcomes in the areas of mathematics and science, particularly for female and minority students. In the language of my own work, "skill," "will," and "social support" components need to be integrated with learning content and context (McCombs & Marzano, 1990). That is, a learner-centered perspective is one that reciprocally focuses on promoting the motivation and higher level thinking of the students and the adults who interact with them. The result is an enabling interpersonal context for the empowerment of natural self-esteem and motivation to learn (will) and the acquisition of cognitive and metacognitive competencies (skill) through quality relationships and interactions with others (social support).

Attention to the larger context of learner and environmental variables in enhancing student achievement in mathematics and science is echoed in current research. For example, Stipek and Gralinski (1991) researched gender differences in achievement-related beliefs and emotional responses to success and failure in mathematics for large samples of lower elementary and junior high school students. Results verified the importance of learners' achievement-related beliefs (attributing failure to ability, expecting to do less well, having low ratings of ability) for explaining gender differences in mathematics performance and in choices of future coursework and occupations. Thus, understanding how students' internal constructions and belief systems can significantly affect performance variables, independent of course content or of teaching strategies, points to the importance of attending to learner-centered principles. In addition, research by Wood, Cobb, and Yackel (1991) highlights the importance of teacher belief systems regarding mathematics and the role of the teacher in initiating and guiding students' development of knowledge (as compared with transmitting information) in enhancing student performance. Because teachers can be reciprocally empowered within transformed models of teaching and of learning, they will be able to focus both on the content of the models (what kinds of problems to teach) and on the students (what strategies best enhance learning and motivation).

Learner-Centered Mathematics and Science Interventions in School Settings

In the area of mathematics, boring and repetitive curricula often are blamed for both the emotional withdrawal and the performance decrements among girls, minority students, and many successful students of all races and both genders. Even though the NCTM has argued that mathematics is a useful, exciting, and creative area of study that should be appreciated and enjoyed throughout students' middle- and secondary-school years, instructional approaches and curricula that take advantage of the students' natural curiosity and motivation to investigate and to understand the world around them are the exception rather than the rule.

Comprehensive School Mathematics Program

One of these exceptions is the Comprehensive School Mathematics Program (CSMP) produced by researchers at the Mid-continent Regional Educational Laboratory (McREL) for kindergarten through sixth-grade students and its middle-school counterpart currently under development, the Comprehensive Middle School Mathematics Program (CMSMP). In contrast to many other existing or redesigned mathematics curricula, CSMP and CMSMP focus on problem solving and communication with attention to self, metacognitive, cognitive, affective, and behavioral dimensions of learning. This curriculum has a situation-based philosophy that makes a minimum of assumptions about the life-styles and experiences of students. Strategies for adapting to individual differences are embedded in tailored teacher training and resource guides that suggest appropriate examples, games, stories, and problems. The specific principles embodied in the CSMP and CMSMP curricula include the following:

- *Every child needs and can learn mathematical skills and concepts.* The curriculum assumes that mathematical processes and principles are of sufficient interest and widespread need that all children can and will acquire them if they are motivated and encouraged within a supportive and creative environment. Furthermore, it is assumed that learning takes place when children react to interesting situations, whether real life or fantasy. These situations involve the students personally and allow arithmetic to take the form of "adventures in the world of numbers."
- *Students should not be denied access to a particular topic because they have not mastered some previous topic.* The curriculum does not build understanding of mathematics concepts and manipula-

tions in a rigid hierarchical fashion and, thus, does not require mastery of prior concepts, thereby allowing all students access to all topics. Students can choose how many solutions to find for each problem depending on their ability and on how interesting the problem is to them.

- *Most students will learn mathematics better if they construct their own understanding (i.e., can write or verbally communicate about mathematics).* Not only are reading, writing, and oral assignments incorporated into the teaching materials, but opportunities for peer teaching and learning in cooperative group interactions are also provided as teaching and learning strategies that help students construct their own understanding.

- *Applications of mathematics should precede the introduction of specific mathematical topics.* The curriculum has maximum flexibility in integrating mathematics applications within other curriculum areas. For example, if a problem determining the authorship of a novel requires a study of elementary statistics, then elementary statistics becomes a topic for study. Because prior knowledge is not prerequisite for any suitable application, most of the mathematics required for an application is developed as needed.

- *A general problem-solving strategy should be used as the focus for learning how to solve mathematical (and other) problems.* Polya's (1969) four-step strategy involves (a) understanding of the problem, (b) constructing a plan to attack the problem, (c) carrying out the plan, and (d) evaluating the results. In applying this strategy, students often are encouraged to search for ways to simplify the problem, thereby gaining an understanding of the solution process for more complex problems.

- *A situational spiral approach facilitates learning mathematics for the student.* It is assumed that learning is more of a spiral process than a straight-line or linear process. That is, intuitive leaps can play as big a role in transforming knowledge structures and understanding as acquiring small successive pieces of information. Thus, every child can and does learn something from each problem-solving situation and its interrelated experiences, sometimes suddenly and dramatically and sometimes latently. By continuously returning to similar problems and concepts throughout the curriculum, students are allowed to assimilate and internalize the concepts and knowledge through repeated, expanded learning situations, thereby enhancing their depth of understanding through changing situations that require synthesis and application of knowledge, not simply a recall of previously mastered facts.

In general, these curriculum materials emphasize student creativity and teacher ingenuity. They also emphasize equity, with all students working together rather than in ability groups. Because the situations are open ended, all students can participate in a meaningful way and learn from each other's reactions. The teacher materials encourage dialogue and questioning. Students are led by their reactions to situations to discover important ideas in mathematics and to think about situations mathematically.

Evaluation data (Heidema, 1991) from more than 200 classrooms indicate that (a) CSMP students are better able than non-CSMP students to apply the mathematics that they have learned to new problem situations; (b) CSMP students learn the traditional mathematics skills and concepts at least as well as comparable non-CSMP students; (c) CSMP students show a higher level of enthusiasm and interest in their mathematics program than do comparable students in more traditional programs; (d) evaluation results are consistent whether one is comparing differences in CSMP students with non-CSMP students by race, gender, or ability levels; (e) CSMP students are more willing than non-CSMP students to tackle unfamiliar problems; and (f) CSMP-trained teachers report having both more confidence and less mathematics anxiety.

Revisions currently being made to the CSMP program and to the CMSMP under development focus on concerns within the larger context of school redesign. From a learner-centered perspective (McCombs, 1992) and a living-systems framework (Hutchins, 1989, 1990), these revisions acknowledge changing paradigms of schooling, curriculum, and instruction as well as changing knowledge and skill requirements to prepare students adequately for the needs of the 21st century. Revision goals include

- introducing students to a broad range of mathematically important and powerful ideas;
- presenting mathematics as a unified whole with interdisciplinary integration;
- encouraging a three-level approach to learning mathematics (understanding the content and applications, developing the techniques and processes for learning that content, and applying the appropriate means in the solutions of problems);
- developing the mathematics concepts through a variety of sequenced, interrelated problem-solving experiences;
- relating the mathematics content to students' natural interests and experiences as well as to future education, work, and leisure goals;
- allowing students to explore, construct, and communicate their understanding of mathematical concepts;
- engaging students actively both intellectually and physically in the learning process; and
- developing students who value mathematics and who are confident in their mathematical abilities.

Dimensions of Learning

These goals are also being addressed within the context of the *dimensions of learning* paradigm for curriculum, instruction, and assessment developed by Robert Marzano and his colleagues (Marzano, 1991; Marzano & Kendall, 1991). This framework specifies characteristics of authentic learning tasks that are both maximally engaging and facilitative of higher order thinking and learning.

Eight characteristics of authentic tasks as identified by Marzano and Kendall (1991) include the following:

- *Student directed.* The task allows for a maximum amount of student control and regulation in terms of its design. It is structured to promote students' acceptance of personal responsibility for their actions through exercising personal choice and decision making within developmentally appropriate task and curriculum structures. Options for student choice might include how to structure or to define the task, what information and resources to utilize, the format of the outcome, and with whom to work. Exercising these choices develops student competence, self-control, and self-discipline while, at the same time, nurturing their interest and motivation to learn.

- *Personally, socially, and domain relevant.* The task is perceived by the learner as falling within his or her set of personal goals, needs, and interests. The task also is considered relevant to broader social goals or issues such as world peace, environmental protection, or human rights. Finally, the task is considered important within the accepted domain of study and falls within content areas considered important to student achievement at the national, state, or local levels.

- *Production oriented.* The task uses some knowledge not currently in long-term memory, and new information must be generated to create new knowledge by the reorganization or manipulation of existing knowledge in long-term memory. It allows for multiple and varied products such as oral reports, written reports, panel discussions, videotaped documentaries, and dramatic presentations.

- *Multidimensional.* The task uses a diversity of cognitive operations that realistically characterize real-world problems. These complex reasoning processes include comparison, classification, structural analysis, supported induction, supported deduction, error analysis, constructing support, extension, decision making, investigation, systems analysis, problem solving, experimental inquiry, and invention.

- *Interdisciplinary.* The task involves knowledge from two or more domains and extends into content domains in ways comparable to real-life tasks. It not only uses concepts, generalizations, skills, and processes that are considered critical to specific content areas, but also combines concepts, generalizations, skills, and processes from two or more content domains.

- *Nonroutine and data based.* The task requires a certain amount of decision making and conscious thought with the execution of component mental processes, regardless of how many times these processes have been used. It also requires the learner to collect and assemble information from a number of sources, some of which are primary sources (e.g., reading, interviewing, making observations, using computerized databases).

- *Long term.* The task is not bound by traditional classroom time structures and approximates upper limits comparable to naturally occurring problems. In the classroom setting, task length could range from one week to one year.
- *Interactional.* The task is highly amenable to cooperative and collaborative work such that it reflects the manner is which tasks are performed in real life. Although tasks can be, and sometimes should be, performed alone, authentic tasks may also be systematically performed by groups of students in the classroom.

Essentially, these characteristics define tasks that allow for a maximum amount of student control and self-regulation, as well as tasks that are amenable to cooperative and collaborative work. Marzano's (1991) model assesses the degree to which tasks are authentic by assessing student abilities in six areas: (a) ability to use concepts, generalizations, skills, and processes considered critical to specific content areas; (b) ability to use complex reasoning processes; (c) ability to gather information in a variety of ways from a variety of sources; (d) ability to create a variety of products; (e) ability to regulate one's own learning; and (f) ability to work in a cooperative or collaborative manner. As students progress through a school system, Marzano recommended that they be assessed by a number of teachers on tasks that cover all six areas. These multiple assessments then could be aggregated, and a cumulative database of each student's performance on the six areas constructed. At any point in time, an up-to-date student report on the number and level of assessments completed in each area could be obtained.

Learner Centered as Reciprocal Empowerment

The learner-centered focus of this work is shared by others who are concerned about reforms in science curriculum and assessment and who emphasize the balance of learner-centered and curriculum or content concerns. For example, Shavelson, Carey, and Webb (1990) emphasized that current reforms must attend to cognitive fidelity and to process relevance. By this, they mean that curriculum and assessment approaches need to focus on the degree to which students conceptually understand content based on their unique interpretations and cognitive constructions. Also, focus must be on how students' unique understandings translate into their ability to apply successfully the concepts and skills deemed important in a field.

Conceptual understanding does not automatically translate into higher level skills for representing and solving mathematical or scientific problems, thereby necessitating explicit instruction and assessment of understanding and of application of this understanding. It also means that students can be assisted by explicit instruction in ways to enhance metacognitive development in such areas as self-control (planning, goal setting, strategy selection, etc.), self-monitoring (of mood, comprehension, learning progress, etc.), and self-evaluation (of progress, strategy use, performance, etc.) as well as reflective self-awareness of self-as-agent in orchestrating cognitive and metacognitive processes (McCombs, 1991). My own work has emphasized this focus in the form of generalized strategy and skill training programs addressing self-development needs for early and late adolescent students (McCombs, 1988, 1989, 1990).

The trainings for students and teachers, within the context of my work with the reciprocal empowerment model, are designed to enhance not only students' development of their own metacognitive self-awareness of their agency, but also teachers' empowerment by realizing their own agency and creativity in seeing how they can nurture and enhance that type of self-development in their students. Students at middle- and high school levels are taught principles of psychological functioning, including their inherent capacity for self-esteem, motivation, and higher level thinking; their agency in creating thoughts and belief systems; their capacity for a higher level of understanding about their agency in overriding constructed thoughts and belief systems; and the relationships among thoughts, feelings, and behavior. Experiential exercises help students to deepen their understanding of personal agency in the thought–feeling–behavior cycle and in accessing higher levels of awareness of inherent capacities for self-regulation, insight, and creativity.

For teachers, the training is presented in a workshop format in which they are similarly introduced to principles of psychological functioning, their agency, and inherent capacities for higher level thinking. They also are exposed to didactic information in areas such as adolescent development, needs of at-risk youth, and effective intervention strategies. Opportunities are provided for teachers to role play, model effective behaviors, and participate in simulated listening and interpersonal exercises. As a result of these experiences, teachers acquire a deeper understanding of students' inherent potentials and agency and, thus, are more able to elicit students' natural motivation and capacity for higher level awareness of their own agency in learning contexts. Finally, the importance of an environment of socioemotional support for students' development of self-regulated learning capabilities is emphasized. Teachers learn the importance of students' positive self-development of close relationships with caring, supportive adults, and of opportunities to participate in decisions that relate to their concerns, interests, and goals within the context of their learning experiences. Teachers realize that allowing students choices and personal control facilitates the development of perceptions of competence that can enhance natural tendencies to want to learn and to do well.

In summary, our collective work at the McREL has been directed at systemic, learner-centered approaches to school redesign at the learning, instructional, and administrative levels of the educational system. Through our partnership with the American Psychological Association and collaborative work on the learner-centered principles, new and exciting learner-centered paradigms are emerging. A real key to the success of these paradigms, however, is the use of powerful strategies for communicating and disseminating the paradigms to educators and to policymakers at local, state, regional, and national levels.

Implications for Further Research and Development

New models and paradigms for reform in mathematics and science education must expand beyond concerns regarding content and teaching strategies. In the absence of a learner-centered focus, significant changes in student achievement are not likely to occur. As the director of our laboratory, Larry Hutchins, has argued, to accomplish the national goals set for mathematics and science education requires a new conception of schooling that is based on the individual needs of students and on a recognition that every child marches to his or her own vision. Furthermore, he argues that standard setting will do nothing until every parent, teacher, and child sees his or her own potential and aspires to greater things. We can think about the symbols that are characterized in mathematics and science content (e.g., data, numbers, words, and oral and visual representations) in a standardized way, not about the *people* who manipulate these symbols (Hutchins, 1991). One of the biggest challenges, then, is to change the belief systems about the nature of the learner, the learning process, and the entire enterprise of schooling. This is not as ambitious as it might sound. We have the technology to support this dream, and a growing number of people are ready for the change.

So what will it take? We believe that students, parents, teachers, administrators, and policymakers can be supported in creating a new vision. The support needs to be in the form of clear examples of what is possible and of how these visions can be accomplished, combined with sound data on the effectiveness of new approaches. Thus, I would recommend that important new directions for further research and development be explored in the following three areas:

- Research and development of alternative strategies for packaging (e.g., via video technology) "paradigm demonstrations" in ways that can create new visions of what is possible and alternative models or designs at all levels of the educational system (instruction, curriculum, assessment, administration). These demonstrations need to be evaluated systematically against changes in the conceptions and visions of teachers, administrators, and policymakers regarding the structure and process of schooling.

- Research and development of new content paradigms in mathematics and science that fully integrate learner-centered principles into curriculum and instruction. These developmentally appropriate curricula, addressing the needs of the whole child, as well as of learner-centered teaching and instructional practices that accompany these curricula, need to be evaluated systematically against cognitive, metacognitive, affective, motivational, and behavioral outcomes and standards.

- Research and development of truly learner-centered strategies that more intensely involve students themselves in school redesign efforts. These strategies need to give to learners (a) more opportunities to exercise control, (b) a major role in educational decision-making promises to develop more responsible learners and citizens, and (c) enhanced motivation to learn and to attain higher levels of educational achievement.

The realization of these goals in mathematics and science education requires attention to strategies for transforming traditional assumptions and perspectives in order to arrive at a new paradigm for our educational system. Because results count, the emerging evidence in support of a new learner-centered design needs to be expanded with carefully defined research on parameters of this design. As evidence accumulates, changes will occur. Then, our national vision of active, self-directed, and life-long learners who are first in the world in mathematics and science achievement by the year 2000 can become a reality.

References

American Association for the Advancement of Science. (1989). *Project 2061: Science for all Americans*. Washington, DC: Author.

Callahan, P., Pogrow, S., & Gore, K. (1990). *Top-rated thinking-in-content curricula for educationally at-risk middle school students*. Tucson: Middle School Curriculum Review Series, University of Arizona.

Covington, M. V. (1991, August). *Motivation, self-worth, and the myth of intensification*. Paper presented at the 99th Annual Convention of the American Psychological Association, San Francisco.

Eisner, E. W. (1991). What really counts in schools. *Educational Leadership, 48*(3), 10–17.

Gore, K., & Pogrow, S. (1991). *Top-rated thinking-in-content curricula for all middle school students*. Tucson: Middle School Curriculum Review Series, University of Arizona.

Heidema, C. (1991). *Comprehensive middle school mathematics curriculum.* Aurora, CO: Mid-continent Regional Educational Laboratory.

Heshusius, L. (1991). Curriculum-based assessment and direct instruction: Critical reflections on fundamental assumptions. *Exceptional Children, 57,* 315–328.

Hutchins, C. L. (1989, August). *Systemic redesign in the educational arena.* Paper presented at the 33rd Annual Meeting of the International Society for General Systems Research, London.

Hutchins, C. L. (1990, Fall). Changing society spells school redesign. *Noteworthy,* 47–48.

Hutchins, C. L. (1991). *Visionary American schools.* Aurora, CO: Mid-continent Regional Educational Laboratory.

Levin, H. M. (1991, April). *Building school capacity for effective teacher empowerment: Applications to elementary schools with at-risk students.* Paper presented at the annual meeting of the American Educational Research Association, Chicago.

Loria, O. C. (1989). *Survey of alternative schools: Keys to successful school redesign.* Aurora, CO: Mid-continent Regional Educational Laboratory.

Marzano, R. J. (1991). *Dimensions of learning: A new paradigm for curriculum, instruction and assessment.* Aurora, CO: Mid-continent Regional Educational Laboratory.

Marzano, R. J., Brandt, R. S., Hughes, C. S., Jones, B. F., Presseisen, B. Z., Rankin, S. C., & Suhor, C. (1988). *Dimensions of thinking: A framework for curriculum and instruction.* Alexandria, VA: Association for Supervision and Curriculum Development.

Marzano, R. J., & Kendall, J. S. (1991). *A model continuum of authentic tasks and their assessment.* Aurora, CO: Mid-continent Regional Educational Laboratory.

Mathematical Sciences Education Board. (1990). *Reshaping school mathematics: A philosophy and framework for curriculum.* Washington, DC: National Academy Press.

McCombs, B. L. (1988). Motivational skills training: Combining metacognitive, cognitive, and affective learning strategies. In C. E. Weinstein, E. T. Goetz, & P. A. Alexander (Eds.), *Learning and study strategies: Issues in assessment, instruction, and evaluation* (pp. 141–169). San Diego, CA: Academic Press.

McCombs, B. L. (1989). Self-regulated learning and academic achievement: A phenomenological view. In B. J. Zimmerman & D. H. Schunk (Eds.), *Self-regulated learning and academic achievement: Theory, research, and practice* (pp. 51–82). New York: Springer-Verlag.

McCombs, B. L. (1990). *Reciprocal empowerment: The key to excellence in education.* Unpublished manuscript.

McCombs, B. L. (1992, August). *Learner-centered psychological principles: Guidelines for school redesign and reform* (rev. ed.). Washington, DC: American Psychological Association, APA Task Force on Psychology in Education.

McCombs, B. L. (1991). Overview: Where have we been and where are we going in understanding human motivation. In B. L. McCombs (Ed.), *Unraveling motivation:*

New perspectives from research and practice [Special issue]. *Journal of Experimental Education, 60,* 5–14.

McCombs, B. L., & Marzano, R. J. (1990). Putting the self in self-regulated learning: The self as agent in integrating will and skill. *Educational Psychologist, 25,* 51–69.

National Council of Teachers of Mathematics. (1989). *Curriculum and evaluation standards for school mathematics.* Reston, VA: Author.

Polya, G. (1969). *Mathematics and plausible reasoning* (Vol. 2, rev. ed.). Princeton, NJ: Princeton University Press.

Shavelson, R. J., Carey, N. B., & Webb, N. M. (1990). Indicators of science achievement: Options for a powerful policy instrument. *Phi Delta Kappan, 71,* 692–697.

Skrtic, T. M. (1991). The special education paradox: Equity as the way to excellence. *Harvard Educational Review, 61,* 148–206.

Smith, G. R., Gregory, T. B., & Pugh, R. C. (1981). Meeting student needs: Evidence for the superiority of alternative schools. *Phi Delta Kappan, 62,* 611–618.

Stipek, D. J., & Gralinski, J. H. (1991). Gender differences in children's achievement-related beliefs and emotional responses to success and failure in mathematics. *Journal of Educational Psychology, 83,* 361–371.

Wood, T., Cobb, P., & Yackel, E. (1991). Change in teaching mathematics: A case study. *American Educational Research Journal, 28,* 587–616.

CHAPTER 10

Developmental Diversity in Mathematical and Scientific Competence

Daniel P. Keating

The ability to work effectively in mathematical, scientific, and technical domains is increasingly important for both individuals and societies. The restriction of career opportunities for individuals who lack these competencies has become greater as a consequence of the information revolution, and the ability of societies to compete in the global economy has grown more dependent on successful innovation, especially technical innovation. Although the attainment of literacy has long been a hallmark of the effectiveness of educational systems—and legitimately so—it is only in recent years that the widespread attainment of at least minimal levels of competence in mathematical and scientific domains has been seen as equally important.

One effective method for discovering the critical features that contribute to the development of competence is to explore the diversity of attainment among individuals and among groups and then to trace the various developmental histories that are associated with that diversity. It has been known for a long time that there is considerable variability of attainment in these domains, from virtual innumeracy to extraordinary accomplishment (Keating, 1976, 1990a; Stanley, Keating, & Fox, 1974). In most instances, these differences have been documented using psychometric methods, with relatively less attention paid to uncovering the developmental histories that might help explain them. The increased emphasis on the importance of such competencies, however, requires understanding as much as possible about their formation to plan more effective instruction and to guide interventions.

To aid in this task, there is a wealth of information that has been accumulated over many years of psychological research from a variety of perspectives within the discipline. Making effective use of this material, however, requires that it be considered critically and that it be approached from a clear and consistent developmental perspective.

In this chapter, I examine several significant lines of research from which to draw to begin constructing a coherent story of the development of mathematical and scientific competence, and suggest features that should be saved from each perspective. Then, I propose several key ideas derived from this critical review, which serve as supports for a revised and more comprehensive model of development, one that can accommodate observations about variability. To illustrate this approach, I examine specific areas of diversity using this developmental perspective and consider some practical implications of the proposed developmental scenario. In other words, given what is known about the development of competence in the mathematical and scientific domains from accumulated psychological research, what should be done to foster their acquisition?

Psychological Approaches to the Understanding of Diversity

The existence of substantial differences among individuals in mathematical and scientific competence has been well and widely documented. There is far less consensus, however, on how best to understand these differences. A brief examination of the strengths and limitations of four major approaches that have proposed different explanations will permit clearer specification of the elements that are valuable in a developmental integration of psychological research. A more detailed examination of these theoretical approaches is available elsewhere (Keating, 1990c, and references therein); in this chapter, I provide only an overview of the key ideas from psychometric, Piagetian, information-processing, and Vygotskian perspectives.

Most attempts to explain individual differences in performance on mathematical and scientific tests and tasks have focused on the characteristics of good versus poor performers. For psychometric theory, the critical features are the mental abilities. In Piagetian theory, they are the logical operations. In information-processing research, the key elements are the specific cognitive processes. In contrast, the developmental approach focuses not just on individual performance, but on how ontogeny (i.e., individual development) proceeds in the context of, and in interaction with, a cognitive socialization environment. The attempt to understand diversity must therefore concern itself fundamentally with developmental histories. Individual differences in performance are a part of developmental diversity, but not the full story of it.

Individual Differences: The Psychometric Model

Historically, the earliest scientific examination of individual differences in abilities was the psychometric approach, beginning at the end of the 19th century with the work of Galton and Binet, and blossoming in the early 20th century through the influential synthesis advanced by Terman (Keating, 1990b). It remains the most systematic approach to the categorization of individual differences and has had the greatest impact on educational practice of any line of psychological research through its practical program of ability and achievement testing. In fact, the magnitude of its impact has itself become a cause for concern among many critics (Frederiksen, 1984), in that it has shaped the kind of knowledge and skills, and hence educational practices, that are considered legitimate.

Despite the broad range of criticism to which the psychometric approach has been subjected, much of which is well founded (Ceci, 1990; Gardner, 1983; Keating, 1984, 1990b; Sternberg, 1990), there are two fundamental discoveries for which it can claim credit. The first is the extraordinarily robust pattern of covariance structures of ability and achievement within and between populations. This degree of consistency undoubtedly indexes fundamental aspects of human intelligence. The second, actually a corollary of the first, is the recognition that population diversity, which follows regular patterns, is a basic feature of human cognitive performance.

Four important aspects of psychometric regularity that characterize the observed diversity are particularly relevant here. First are the moderate-to-high, but always positive, correlations across ability and achievement areas. Although the long-standing battle between general versus specific psychometric abilities has moved into cognitive science as a conflict between domain-general learning versus domain-specific expertise (Keating & Crane, 1990), the reality is that there is overwhelming evidence for both. Because the current zeitgeist is to emphasize domain specificity (with good reason, given the earlier dominance of general ability models), it is important not to lose sight of the fact that individuals who do well in school and on psychometric tests tend to do so across the board, and most students who do poorly also do so broadly. This implies a significant constraint on theories that would seek to account for differences in cognitive performance, namely, that they need to identify important general factors that are formative in the development of abilities.

The second finding is that, embedded within this pattern of general ability covariance are strong gradients of performance associated with group membership. The three that have been studied most extensively are gender, social class, and race or ethnicity. For a variety of historical reasons (cf. Keating, 1984, 1990b), the psychometric approach to group differences has been less than welcome in many quarters of society. Some scientists (and nonscientists) have too readily adopted an a priori stance that such differences necessarily imply deficits, and have advanced educational and social policies without paying sufficient attention to the complex contextual history out of which such differences emerge (Ceci, 1990; Gould, 1981). Comprehensive developmental accounts necessarily address the contexts of development, and such factors need to be prominent in attempts to explain diversity. For the moment, it is important not to deny that such gradients exist (although their magnitude and shape merit careful analysis; e.g., Hyde, chapter 7, this volume), but to pursue the research that would permit the telling of a complete developmental story rather than to accept superficial explanations that rely on simple, main effect explanations.

The third finding that must be included in any complete story of ability development is the intraindividual consistency of test and school performance across ages. This consistency is, of course, greater over one to several years and decreases somewhat over longer time periods. Although it is important to recognize individual exceptions to this pattern, the impact of early school success or difficulty is quite strong. This may be especially true for mathematics, in which the cumulative nature of expertise may be particularly strong (Case, 1992).

The fourth key aspect is the counterpart to the first: The substantial degree of domain specificity that is apparent as one examines cognitive performance more closely (Gardner, 1983). Within the psychometric model, this tension between general and specific abilities has been resolved largely by the compromise solution of hierarchical models (Anastasi, 1990). But such a solution provides more of an organization for these competing tendencies than an explanation of them. From a cognitive science perspective, Sternberg (1989) contended that the argument between domain-general and domain-specific claims was fruitless. So long as the argument remains at the level of how to characterize best the organization of expertise or, in other words, the *products* of intellectual activity, then indeed it is unresolvable. However, if we reconstruct the question in terms of the developmental processes that give rise to both domain-general habits of mind and domain-specific expertise, then we can generate a productive theoretical and practical research agenda (Keating, 1990c; Keating & Crane, 1990).

The principal limitation of a purely psychometric approach is that, in focusing on the products of intellectual activity, it has been unsuccessful in generating a comprehensive model of the relationship between such products and the processes that yield them (Keating & MacLean, 1987). To restate this more positively, psychometric research has produced robust and regular patterns that capture important structural features of inter- and intraindividual diversity of performance. It is useful to regard these regularities (general ability, group gradients, consistency across age, and domain specificity) as targets of explanation that any viable process model must be able to meet.

States of Logic: The Piagetian Model

Although the Piagetian approach has focused much more on universal regularities in development than on diversity, it is an important model that can contribute significantly to our understanding of diversity. This is particularly so regarding diversity in mathematical and scientific thinking, given its central focus on logical and mathematical structures and its explicitly developmental approach (Keating, 1990a, 1990d). A central goal of the Piagetian approach has been to identify the underlying developmental shifts that can account for observed changes in cognitive performance. In this regard, it is conceptually complementary to the psychometric model and an important component for any potential developmental synthesis.

Two features of the Piagetian (and neo-Piagetian) model are especially salient. The first is the notion of logically organized cognitive structures that set the pattern for a wide range of related intellectual activities. In this regard, Case's (1992) central conceptual structures were a productive reworking of this core idea. In contrast to some "hard-stage" interpretations of the Piagetian canon, central conceptual structures are broad but not necessarily all-encompassing. At critical points in development, previously localized shifts in the understanding of related phenomena coalesce or cascade into a much broader cognitive reorganization. Among most 4- to 6-year-olds, one such reorganization is the consolidation of the notion of a number line that incorporates ideas about quantity, order, seriation, and simple addition and subtraction. The acquisition of such a central conceptual structure has profound implications for how the child views the world in a variety of domains, some more directly than others. In contrast, the failure to acquire such a notion may leave the child incapable of benefiting from most standard instruction in the early years of schooling, particularly in mathematics.

The second major feature is the concept of equilibration as a core metaphor for change, that is, the creative tension between accommodation and assimilation. Most research activity using Piaget as the starting point has focused on the nature of logical stages and their presence or absence as explanatory. Until recently, there has been less effort given to understanding the mechanisms of change themselves (Inhelder & de Caprona, 1990). Updating these notions to integrate them with contemporary work on iterative feedback functions and their role in self-organizing dynamic systems is a potentially valuable route for examining developmental, rather than just logical, processes (Keating, 1990c).

Current work on mathematical and scientific misconceptions (or naive conceptions) becomes more forceful from such a perspective. Conceptual growth is more than the gradual accumulation of skills. Instead, it competes with existing notions of how the world works, notions that have been built up through recursive interactions of the learner's experiences and of the cognitive structuring of those experiences. In many cases, learning a new concept requires giving up a scheme that had worked pretty well in the past. For example, the early adolescent's acceptance of the notion of logical uncertainty seems to have few advantages initially over the simpler but incomplete logical system of childhood (Keating, 1990a, 1990d).

Similarly, we found that many university mathematics students were unable to apply the appropriate proportionality algorithms to simple concrete problems, suggesting that different representational systems of the same formal domain can coexist within the same individual and that the integration of intuitive notions of proportionality with formal algorithms is far from automatic (Keating & Crane, 1990). Focusing on the developmental mechanisms through which change occurs and recognizing the tension between the desire to retain existing schemes and the need to accommodate conflicting information are central components of a developmental understanding of diversity (Gardner, 1991). For a host of reasons, social and emotional as well as cognitive, individuals differ in the degree to which they attend to disconfirming information and in the degree to which they are made uncomfortable by it (Kuhn, Amsel, & O'Loughlin, 1988; Shafrir, Ogilvie, & Bryson, 1990). Thus, differences in style as well as timing may play a crucial role in the emergence of diversity.

Aspects of the Piagetian approach that have been most frequently criticized and least supported empirically, such as the universality of logical stages, the elevation of a particular (i.e., propositional) logic as an end goal of development, and the preeminence of logical above other cognitive changes, are real limitations of the model. I have argued that these reflect the creative tension between a desire for closed structures in conflict with a recognition of open systems (Keating, 1990d). Nevertheless, by integrating the more crucial features of the Piagetian approach (central conceptual structures and dynamic equilibration that implies both gain and loss in development) into a more comprehensive developmental model, understanding of the origins and growth of human intelligence will be enhanced.

Computational Systems: The Information-Processing Model

In more recent years, psychologists have attempted to create more precise models of human cognitive performance by using a wide range of techniques enabled by technological advances, especially high-speed computing. In earlier versions, these were termed *information-processing models* because they consciously adopted the computer as a metaphor for the human mind. More recently, these approaches have begun to incorporate complex simulations as well as emerging evidence from the neurosciences, and have adopted the broader terminology of cognitive science.

For the purposes of studying developmental diversity, the information-processing approach shares important common assumptions with the Piagetian model. Most notably, the shared goal is to understand the cognitive processes that underlie intellectual performance, rather than to categorize the intellectual products themselves. But there are also important differences between Piagetian and information-processing approaches. From the latter perspective, no centrality is given to shifts in the structure of logic over any other cognitive changes. Instead, the goal is to characterize, as completely as possible, the cognitive activity occurring in real time as individuals engage in intellectual tasks.

Sternberg (1985) proposed a componential approach to accomplish this goal. He sought to decompose complex cognitive tasks using experimental and statistical methods of partialing out the observed variance on those tasks. Siegler (chapter 11) used both reaction time and observation of strategies in early arithmetic computation to identify important benchmarks in the growth of that performance. I have explored some of the real time cognitive processing correlates of performance on both general and specific ability tasks (Keating & Bobbitt, 1978; Keating, List, & Merriman, 1985; Keating & MacLean, 1987). Many other researchers have explored similar issues using a range of experimental and observational methods.

Although it is premature to summarize this highly active field of psychological research, some important lessons for the understanding of developmental diversity have already emerged (Ceci, 1990; Keating, 1990c; Sternberg, 1990). The first of these is that we have been unable to identify independent cognitive entities or processes that are in themselves sufficient to account for the observed variability of performance. We have, however, generated a long list of important cognitive activities that vary across individuals. Variance among persons in these activities (on some tasks and on some occasions) covaries with performance on complex, real-world tasks. A provisional list would include speed of processing; the availability and accessibility of appropriate procedural strategies; the magnitude, organization, and relevance of specific knowledge and information; the conceptual structures that organize procedural and declarative knowledge; and self-regulatory activities, including metacognition and attentional habits. This rough categorization is unlikely to be an exhaustive list.

There are several ways of regarding this burgeoning complexity. The first, perhaps, is to accept, with regret, the fact that we are yet again unable to identify the key elements of cognitive processing that underlie important intellectual performances. A second is to view this research agenda as incomplete, but promising. One might then pursue one or several parameters of cognitive processing, with the expectation that they will prove to be the critical features explaining cognitive performance. Both of these views converge on the belief that there is one (or a small number) of cognitive processing elements that can account for the variability in intellectual performance and that separating these central features from more peripheral ones is the key to understanding.

The third approach begins with the assumption that the substantial empirical work implicating a wide range of cognitive processes has, in fact, yielded valid and substantial relationships. In other words, each of the cognitive activities noted earlier does matter, at least at some point in the acquisition of the expertise that we measure as intellectual products.

If we accept this general principle (reserving the right to judge any particular research finding on its own merits, of course), then we are faced with additional challenges. The degree of complexity introduced by including a long list of determinants of diversity seems initially incompatible with the observed psychometric regularity noted previously. How do these various cognitive processes become integrated so as to yield domain-general and domain-specific coherence, robust gradients by group membership, and across-age consistency?

I propose that this challenge forms the key question for an integrated research agenda with the goal of understanding human diversity (Keating, 1990b). In place of a search for the core elements that form intellectual products, the developmental approach asks how the many different aspects of cognitive activity become integrated over the course of development to yield observed performance patterns (Keating, in press-b). An interesting example from one key domain of how one might proceed is Siegler's attempt (see chapter 11) to look at the role of declarative knowledge (via association strengths), strategies, and self-regulatory behaviors in the acquisition of early arithmetic skills. We might also wonder how those changes articulate with shifts in Case's (1992) central conceptual structures. How, and in what ways, do the changes in children's ability to respond to arithmetic tasks coincide with their conceptual knowledge about numbers? As we proceed with this developmental research agenda, it is likely that numerous opportunities for diversity to emerge will become apparent.

Two immediate implications of such an approach should be noted. The first is that multiple pathways to the development of expertise are likely to exist in any domain, even in mathematics, which appears deceptively simple to model. The tendency in education has been to determine a normative pathway that accommodates the largest number of children (perhaps we should call this the *modal pathway*), and to regard with suspicion deviations from that pathway (Keating, 1990b). Siegler's example of the discouragement of finger counting is pertinent here.

The second implication is an emphasis on the context of acquisition. Contextual effects often have been regarded as the bane of experimental attempts to isolate the crucial cognitive processes that can account for diversity in cognitive performance, and their pervasive recurrence has been a source of considerable gnashing of theoretical teeth. But contextual effects are just another part of the developmental process. Indeed, they offer a meaningful opportunity to identify critical points in the development of expertise because they become incorporated into the pattern of cognitive activity. For university students who show little or no connection between their intuitive notions of proportionality and their sophisticated algorithms dealing with the same topic, we can infer an educational context that affords little direct opportunity to make such connections (Keating & Crane, 1990). In contrast, educational contexts in which extensive discourse about mathematical issues occurs, in preference to an exclusive focus on the acquisition and application of algorithms, appear to be linked to considerable growth in principled, conceptual knowledge.

Contextualism: The Vygotskian Model

The three major lines of psychological research that have been addressed so far begin from a common epistemic perspective, which is that the central focus for the understanding of individual differences is, in fact, the individual. It is the way in which the individual's mental activity is organized, in which that organization changes with development, and in which it differs across individuals that matters. There are alternate epistemic perspectives to this central focus on the subject. One that has become more influential in recent years arises from the work of Vygotsky (1979). Instead of a focus on internal subjective aspects of mental activity, Vygotsky's focus was on how the objective, external conditions of life, especially social conditions, served to shape the human mind.[1]

[1]A more complete epistemic picture would need to include a phenomenological perspective as well, in which no distinction between subjective and objective reality is made, as in John Dewey's pragmatism or J. J. Gibson's ecological perception. This topic is beyond the scope of this chapter, but see Keating (1990c) and Keating and MacLean (1988).

Several important elements in the Vygotskian approach, which dates from the 1920s and 1930s, have recently become more attractive to Western researchers. The first flows from the previous discussion. As contextual effects have become more and more evident, so has the need for a theoretical approach that can accommodate such effects. Vygotsky's notions of how the context of development operates have much to recommend them. A key principle is that thinking originates in external discourse that becomes progressively internalized. Thus, we carry with us structures of knowledge that reflect the social experiences, principally discourse, that were formative in our development. The centrality of discourse for development has been increasingly recognized, both for the social sciences (Newmann, in press) and for mathematics (Stevenson & Stigler, 1991).

A related notion is that for discourse to be useful in advancing development, the discourse has to occur in what Vygotsky (1979) termed the *zone of proximal development*. If it merely repeats knowledge that an individual has already acquired, or that is more advanced than the individual can grasp, then the discourse does not actually involve the individual. Given the likelihood of multiple pathways to expertise noted earlier, and the well-documented diversity of intellectual performance, it is doubtful whether strictly normative mathematics instruction is effective for many students because it fails to engage them in their zone of proximal development (Keating, 1990b, 1991). One final point worth noting is that, although Vygotsky was in strong agreement with Piaget that qualitative mental reorganization was a hallmark of cognitive development, his claim was that such qualitative shifts were more often idiosyncratic than normative or universal. The regularities that are observed would typically be viewed as a function of common or similar experiences, and that deviation from that common history might lead to different outcomes.

Two significant limitations of the Vygotskian approach should be noted. The first is that, until recently, it has not generated much substantiated research, at least by Western standards, that would address the interesting theoretical claims. This may be due more to historical accident than to an inherent weakness of the theory, but such concerns are not insignificant. The second limitation is more strictly theoretical and not as germane to this discussion. Vygotsky (1979) argued that both phylogenetically (i.e., for the species) and ontogenetically (i.e., for the individual), the origin of human intelligence lies in the combination of the use of tools and the use of signs; that is, when we are first able to use language (signs) to guide our practical efforts, then we are engaging in truly human intelligence. The origin of this claim is no doubt familiar to those who are also familiar with the Marxian notion that the origins of human society lie in the social organization of the means of production. The source of an idea does not invalidate it, of course, whatever one's ideological perspective. The specific limitation that should be noted, however, is that this perspective lays claim to instrumental intelligence as paramount over and above other potential claims—for example, the social organization of nurturance as a prelude to human social intelligence and, thus, to culture. Nonetheless, a theoretical perspective from which individual intelligence is seen as originating in and indivisible from social interaction is an important counterweight to the internal, subjective assumptions of most Western psychological research.

Charting Pathways to the Development of Expertise

To advance understanding of the origins of and pathways leading to the observed diversity among individuals in mathematical and scientific competence, the information derived from the perspectives described earlier will need to be integrated. The need for integration of organismic and contextualist models has become a central focus of much of the current work on the development of human intelligence (Ceci, 1990; Keating, 1990c, 1990d; Sternberg, 1990). I have suggested (Keating, 1990c) three elements that are crucial to such attempts at integration.

First, cognitive activities need to replace mental abilities as the basic building blocks for theoretical understanding and for practical assessment. The mental ability construct has proven useful for categorizing empirical observations about diversity, but the too ready reification of these constructs has granted them a causal status that cannot be sustained. If the multiple pathways by which individuals acquire (or fail to acquire) certain competencies are to be understood, then ability constructs must be seen as indicators or markers of competence, rather than as self-evident explanations of competence.

Second, the unproductive categorical dichotomies between achievement and ability, and between general intelligence and specific abilities, must be replaced with the notion that domain-specific expertise and domain-general habits of the mind are developmentally complementary, not competitive. By habits of mind, I mean not only the ways in which individuals attend to and take in information about the world, and in which that information changes their mental schemes and structures, but also a range of "noncognitive" aspects: motivation, effort, aspiration, emotion, and so on (Keating, in press-b). The building up of specific expertise arises from the way in which individuals approach learning, as well as from the way in which the social and cognitive environment is organized. A broad term to cover both notions, and an area to which researchers need to devote more attention, is cognitive socialization.

The third point is a practical one that flows from the first two ideas. Developmental assessment that can inform adaptive instruction needs to replace diagnostic assessment that is used for categorical placement. Because psychologists have tended to focus on the adaptability (or lack thereof) of the child as a hallmark of competence, we need to rethink what we mean by adaptation. Specifically, we need to apply it to educational practice rather than to the child.

Those individuals who give evidence of being best adapted to current social and educational practices, revealed in test scores and school performance, are traditionally defined as most generally adaptable (i.e., intelligent) because of a more optimal underlying design. A consequence of this conflation of two quite different meanings is the assumption that educational difficulty is legitimately explained as a failure of adaptability of the student. From a developmental perspective, it is recognized that success in a particular ecological (i.e., cognitive socialization) niche is not necessarily a sign of adaptability to a wide range of niches. Moreover, a developmental perspective strengthens the tendency to look for ways in which the instructional environment has failed to adapt to the developmental diversity that differential histories inevitably generate. Shifting the onus from a lack of adaptiveness in the child to a lack of adaptiveness in the setting will encourage a close examination of ways to design better learning environments, rather than simply demarcating presumed design flaws in the child. The exploration of some examples of diversity illustrates how the developmental approach permits a more integrated understanding.

Patterns of Competence

One of the more dramatic examples of diversity in mathematical competence is drawn from the Study of Mathematically Precocious Youth (SMPY) that originated at Johns Hopkins University (Benbow & Stanley, 1983; Keating, 1976; Stanley et al., 1974). I draw on the widely reported findings of this study to make a few salient points. First, the diversity that exists even among those students who show the greatest level of competence is considerable. Prior to SMPY, the striking degree of this diversity had not been documented, and we were able to discover it by the conceptually simple expedient of using sufficiently high-level assessments that were capable of revealing higher levels of performance. Second, the competence that these young people displayed on test performance accurately predicted their ability to be equally competent in appropriate educational settings. Students who as early adolescents (mostly aged 11 to 14 years) had test performances that were the equivalent of highly selected high school graduates about to enter university were likely to perform as well in university classes as their older counterparts.

Taken together, these two findings clearly support the need for developmentally sensitive assessments that lead to decisions about how to adapt instruction to meet educational needs (Keating, 1991). There is a wide range of ways in which instruction can be appropriately adapted, but the necessity of doing so is strongly indicated by this evidence. In Vygotskian terms, there is little chance for these students to experience discourse within their zone of proximal development unless the full range of that zone is identified.

Some inferences that have been drawn from these findings, however, appear to be unwarranted. The evidence that high performance on these demanding tests was linked closely to similarly high performance in appropriate (i.e., university) settings does serve to validate the tests in that regard. But it is important not to overgeneralize. The obverse claim, that low test performance is indicative of lack of talent, requires additional evidence that was not the focus of SMPY.

This raises again the issue of multiple developmental pathways. Because variability typically has been explored within a small number of more normative pathways to the development of expertise, diversity or difference has often been equated with deficit. Among early adolescents who have been identified as highly competent in a broad sense, the diversity among them in terms of areas of special expertise is as great as that among unselected groups (Keating, 1991). But so long as consideration is bound to an essentially univariate view, the range of ways in which expertise might develop is unlikely to be explored, and alternative cognitive socialization environments within which such expertise might flourish more readily are unlikely to be created. This is not an idle concern, given the increasing importance of mathematical and scientific competence and the emerging evidence on lagging performance among North American students compared with their Asian and European counterparts.

A second set of premature inferences is the denigration of socialization explanations of the observed diversity in favor of a priori biological differences, especially with regard to gender differences (Benbow, 1988; Benbow & Stanley, 1980). There are two principal flaws in this reasoning. The first is that, by eliminating socialization hypotheses, the credibility of biological origins is enhanced. What is needed instead is a plausible description of the biological pathway by which proposed differences become manifest in the performance in question. The associations among patterns of differences, regardless of how robust or weak such associations may be, provide the starting ground for theoretical claims, not the end point. The second problem is the selection of socialization hypotheses to be discredited. Generally speaking, these have focused on socialization differences that are both weak and too late in development to account for the observed diversity. However, there is a large number of plausible socialization accounts that focus on early experiences that have yet to be explored, as well as accounts that consider these early differences with regard to their interaction with later experiences.

A long-standing criticism of socialization accounts that refer back to experiences in early childhood is that they propose cause–effect relationships that are incommensurate. How could apparently small differences in early childhood manifest themselves as such dramatic differences in performance later in life? Within traditional linear models, this incommensurability does pose a major obstacle. But as the study of dynamic, nonlinear systems in a variety of disciplines whose complexity is similar to that of human development has shown, self-organizing systems can be exquisitely sensitive to initial conditions (Keating, 1990c).

It is important to recognize that the possibility of large effects from small early differences has, at the moment, no more empirical support than the claim of preexistent biological differences. Interestingly, the developmental model has no difficulty in accepting both proposals on equal terms. In both cases it requires a plausible description of the developmental processes and pathways by which such effects become manifest. Indeed, for a developmental model, it is precisely the history of the ongoing interactions between the biological organism (that is restructured continually by its experiences) and the socialization environment (the impact of which is constrained by the current internal state of the organism) that forms the theoretical core. In this sense, it recasts the traditional nature versus nurture controversy into an integrated search for how development proceeds through the interaction of the person with the environment.

Developmental Anomalies

Implementing the developmental agenda described in this chapter requires that researchers employ a range of investigative methods (Gould, 1986; Keating, in press b-c). One method that can be borrowed effectively from the study of biological evolution is the study of anomalies. Within the broad picture of regularity of performance, there are potentially important violations of expectations. Understanding such anomalies can sometimes be more instructive than the observation of regularity, because there are so many plausible explanations for complete regularity. The monotonic growth of skills during childhood, for example, is potentially explicable by growth in speed of processing, declarative knowledge, procedural knowledge, logical understanding, cognitive self-regulation, and so on. Because each increases monotonically, monotonic increase is not a reliable empirical guide for choosing among hypotheses.

Interpretable deviations from regularity, on the other hand, can lead to interesting inferences. We discovered one such anomaly in an investigation of the concept of proportionality (Keating & Crane, 1990). In preparation for some studies of children's understanding of proportionality, we tested some university mathematics students with the intention of using them as pilot subjects. We selected students who were in precalculus courses, and who had passed qualifying tests for those courses. We also administered a paper-and-pencil test of ratios, fractions, and proportions and included only those students who met a mastery criterion of 90%. (Items were drawn from practice tests of the Graduate Record Examinations.) We used these multiple criteria to ensure that our sample of experts had indeed mastered the set of mathematical algorithms associated with proportions.

In contrast, the experimental task we administered was simple. On each trial, the subjects were presented with a pair of geometrically regular wooden objects (circles and squares, larger and smaller in size) that had been divided into segments (thirds, fourths, sixths, and eighths). We then removed one or more segments from each object and asked which object had lost more of its own whole. The mathematical equivalents for these problems were trivially easy compared with the preexperiment criterion test; the most difficult was 3/8 compared with 1/3. We designated some trials as key trials, when a simple visual inspection of angles or areas might lead to a contrary choice to the correct proportional solution, and we used a criterion of 35% errors or greater on these key trials as an indication of less than complete grasp of the proportionality principle. Among this select group, we found that 45% did not consistently display an understanding of proportionality, when such a display required demonstrating transfer to an extremely simplified practical analog.

Equally interesting was the fact that nearly all of these students apparently made their selection on nearly all of the incorrect trials on the basis of an intuitive notion of proportionality. In debriefing immediately after each session, these students explained their incorrect choices as deriving from a visual comparison of relative angles or of relative areas. In other words, they appeared to rely on a developmentally earlier, intuitive notion of relative amounts, rather than to access a well-worn mathematical algorithm that would have rendered the problem quite simple for them. The successful students, of course, explained in debriefing that the problems were quite simple, that they required a simple fractional comparison, and that they had solved them as such.

This pattern of performance among students who were successful by most traditional criteria suggests potentially important inferences about the context in which these focal algorithms were learned. It seems unreasonable to infer that these students lacked the conceptual grasp to handle the problems because they displayed such a grasp on more challenging problems in their coursework and on the pretests. Instead, it seems far more likely that there was little opportunity during their acquisition of these algorithms to link that knowledge to intuitive notions of proportionality. Rather than a conceptually integrated and principled understanding of proportionality, these students seemed to have two relatively independent concepts: an intuitive notion that deals with the real, physical world, and an algorithmic notion that is applied to formal problems. This anomalous performance thus appears to inform us simultaneously about the nature of development and of education. Principled integration of conceptual knowledge is not automatic, and the educational experiences of many successful students appear not to have provoked such integration.

Developing Mathematical and Scientific Habits of Mind

Most of the research on individual differences in mental abilities and educational achievement has proceeded from a *categorical perspective* (Keating, 1990b, 1991) as opposed to a *developmental* one. From this traditional view, a principal goal is to document the essential features that distinguish good from poor cognitive performance. These essential features are typically thought to be internal to the individual.

In this chapter, I have tried to sketch the critical features of a developmental view, one that not only recognizes the valuable empirical evidence on performance differences but that also proposes a more comprehensive explanatory model. The goal of this model is to integrate the analyses of individual patterns of performance with a complementary analysis of the developmental history that contributed to their formation. To emphasize some of these critical features, I have chosen to use the term *developmental diversity* as an alternative to the term *individual differences* for two reasons. It is both more neutral with regard to origins and formative processes, because *developmental* encompasses both biological and socialization effects (and their interactions across time), and is more inherently multivariate, and because *diversity* is more readily perceived as an essential characteristic of any population analysis rather than as a search for deficits. The notion of cognitive socialization is equally central. In the same way that phylogenetic speciation cannot be understood outside the context of the ecological niches in which the speciation has occurred, it is not possible to understand ontogenetic differentiation without reference to the social environment in which the cognitive structures and habits of mind have developed.

The views of development and diversity that I have outlined offer renewed opportunities to understand the sources of cognitive performance differences. Using these perspectives leads to better understanding and integration of the rapidly emerging work in the field, some of the best of which is represented in this volume. For those concerned, however, about the educational contributions of psychological research to education, it is important to consider how these new perspectives can guide approaches to instruction and intervention. Although a comprehensive overview of all the potential educational implications is beyond the scope of this chapter, and is in any case premature given the rapid expansion of our understanding of how mathematical and scientific competence develops, some general principles are already clear.

Some time ago, I suggested that educators needed to expand their views of creativity or innovation and suggested that there were four key aspects of creativity that ought to be of central concern (Keating, 1980): content knowledge, creative thinking, critical thinking, and communication. This framework is particularly appropriate for the domains of mathematical and scientific competence, for which creativity and the capacity for innovation are so crucial.

The first focus was on content knowledge. In 1980, more than today, the necessity of simply having a large amount of knowledge, information, and skills in any domain in which one wished to work was underestimated. It has become increasingly clear through recent research on the importance of knowledge bases in cognitive development (e.g., Chi, Hutchinson, & Robin, 1989; Glaser, 1984) that this is indeed a central factor.

Of course, not all knowledge is equal, and the way in which that knowledge is organized has a major impact on its accessibility and utility, as noted earlier in the discussion of developmental anomalies. The second central feature is critical thinking. By this, I mean the ability and the disposition to seek the underlying relationships among phenomena. (To "cut to the root of the matter" is an etymologically accurate colloquial version of this notion.) Such critical thinking appears to be an important ingredient for the generation of integrated, principle-based conceptual grasp, as opposed to divided and superficial understanding (Keating, in press-a).

In contrast to earlier views that would place them in opposition, creative thinking is closely linked to critical thinking. Creative thinking is far more than ideational fluency or divergent thinking, although it can make use of those strategies when appropriate. More to the point, however, is an appreciation of the provisional and open-ended nature of knowledge, including mathematical knowledge (Kline, 1984). The ability to hold this view without lapsing into complete relativism and skepticism is not easy, and perhaps because of this, most educational practices do not recognize the open-ended nature of knowledge. The tension between criticism and creativity is unavoidable, of course, but it should be viewed as a productive tension rather than as one to avoid. It is important to note that these features are not only, or perhaps even primarily, cognitive. Creative and critical thinking arise from differences in how children approach learning and knowledge. As such, they need to be developed not as yet another set of mechanical skills to be learned, but must be acquired in a context that socializes students to understand their importance (Keating, in press-a).

Increasingly, the role of discourse as a central engine for cognitive socialization has been recognized; thus, the fourth element is communication. This should be seen both as a productive process, in that it affords the opportunity to gain content knowledge and to practice critical and creative thinking, and as a desirable educational goal, in that the purpose of formal cognitive socialization (i.e., education) is to integrate the next generation into the ongoing work of society.

These four features can be reasonably viewed as elements of a mathematical and scientific habit of mind. Much current psychological research (including that reported in this volume) is germane to creating contexts of cognitive socialization that are conducive to the development of such habits of mind. In conclusion, I highlight two that seem to be of the highest priority.

The first arises from the understanding that learning and development are not purely individual phenomena, but are at the same time inherently social. This is evident in a variety of ways already noted; however, the notion of the centrality of discourse among learners and between learners and experts is perhaps the key. For those whose past educational experiences make the notion of mathematical discourse in the classroom an almost unimaginable idea (unless timestable recitations are included as discourse), the emerging evidence from cross-national comparisons and from current work in North America is eye opening (Stevenson & Stigler, 1991). One of the essential aspects of the successful use of discourse in these settings is the skill and knowledge of the in-house expert, that is, the teacher. It is far more difficult and requires a far more sophisticated conceptual understanding to entertain discourse among learners than it is to teach a specific algorithm. One needs to understand enough to find the kernel of correct logic among the chaff of incorrect approaches and to be able to play that back to the group as whole. Even for elementary arithmetic, the possible conceptual depth is considerable (Case, 1992), and students often get by without the conceptual understanding and acquire only the surface.

The second is that an increasing understanding of the pathways along which the development of expertise is possible is leading to a better grasp of two key elements: (a) that the timing of instruction and intervention is at least as important as their content and (b) that there are likely to be multiple pathways to the development of expertise. Both of these notions are especially germane to the development of mathematical competence, which has been traditionally viewed as unidimensional so that if one gets off the normative pathway at any point, alternative routes or later reentry are seen as impossible. (Compare the proliferation of adult literacy training with the virtual absence of programs on adult numeracy.) This is particularly unfortunate for groups for whom the normative pathway has not been traditionally productive.

By expanding their notions of the ways in which mathematical and scientific competence may be achieved, researchers and educators are more likely to find successful routes for a much broader segment of the population. As noted in the introduction, such expansion of competence throughout the population is increasingly necessary for societies that wish to remain active in the global economies of the future.

References

Anastasi, A. (1990, August). *Are there unifying trends in the psychologies of 1990?* Invited address, 98th Annual Convention of the American Psychological Association, Boston.

Benbow, C. P. (1988). Sex differences in mathematical reasoning ability in intellectually talented preadolescents: Their nature, effects, and possible causes. *Behavioral and Brain Sciences, 11*, 169–232.

Benbow, C. P., & Stanley, J. C. (1980). Sex differences in mathematical ability: Fact or artifact? *Science, 210*, 1262–1264.

Benbow, C. P., & Stanley, J. C. (Eds.). (1983). *Academic precocity*. Baltimore: Johns Hopkins University Press.

Case, R. (1992). *The mind's staircase*. Hillsdale, NJ: Erlbaum.

Ceci, S. J. (1990). *On intelligence . . . more or less*. Englewood Cliffs, NJ: Prentice Hall.

Chi, M. T. H., Hutchinson, J. E., & Robin, A. F. (1989). How inferences about novel domain-related concepts can be constrained by structured knowledge. *Merrill–Palmer Quarterly, 35*, 27–62.

Frederiksen, N. (1984). The real test bias: Influences of testing on teaching and learning. *American Psychologist, 39*, 193–202.

Gardner, H. (1983). *Frames of mind: The theory of multiple intelligences*. New York: Basic Books.

Gardner, H. (1991). *The unschooled mind*. New York: Basic Books.

Glaser, R. (1984). Education and thinking: The role of knowledge. *American Psychologist, 39*, 93–104.

Gould, S. J. (1981). *Mismeasure of man*. New York: Norton.

Gould, S. J. (1986). Evolution and the triumph of homology, or why history matters. *American Scientist, 74*, 60–69.

Inhelder, B., & de Caprona, D. (1990). The role and meaning of structures in genetic epistemology. In W. F. Overton (Ed.), *Reasoning, necessity, and logic: Developmental perspectives* (pp. 33–44). Hillsdale, NJ: Erlbaum.

Keating, D. P. (Ed.). (1976). *Intellectual talent: Research and development*. Baltimore: Johns Hopkins University Press.

Keating, D. P. (1980). Four faces of creativity: The continuing plight of the intellectually underserved. *Gifted Child Quarterly, 24*, 56–61.

Keating, D. P. (1984). The emperor's new clothes: The "new look" in intelligence research. In R. Sternberg (Ed.), *Advances in the psychology of human intelligence* (Vol. 2, pp. 1–35). Hillsdale, NJ: Erlbaum.

Keating, D. P. (1990a). Adolescent thinking. In S. Feldman & G. Elliott (Eds.), *At the threshold: The developing adolescent* (pp. 54–89). Cambridge, MA: Harvard University Press.

Keating, D. P. (1990b). Charting pathways to the development of expertise. *Educational Psychologist, 25*, 243–267.

Keating, D. P. (1990c). Developmental processes in the socialization of cognitive structures. *Entwicklung und Lernen: Beitrage zum Symposium anlaBlich des 60. Geburtstages von Wolfgang Edelstein.* [Development and learning: Proceedings of a symposium in honour of Wolfgang Edelstein on his 60th birthday.] Berlin: Max Planck Institut fur Bildungsforschung.

Keating, D. P. (1990d). Structuralism, deconstruction, reconstruction: The limits of reasoning. In W. F. Overton (Ed.), *Reasoning, necessity, and logic: Developmental perspectives* (pp. 299–319). Hillsdale, NJ: Erlbaum.

Keating, D. P. (1991). Curriculum options for the developmentally advanced: A developmental alternative for gifted education. *Education, Exceptionality, Canada, 1*, 53–83.

Keating, D. P. (in press-a). Critical period for critical thinking: The adolescent in school. In F. Miller (Ed.), *Adolescents, schooling, and social policy*. Albany, NY: SUNY Press.

Keating, D. P. (in press-b). Habits of mind: Developmental diversity in competence and coping. In D. K. Detterman (Ed.), *Current topics in human intelligence*. Norwood, NJ: Ablex.

Keating, D. P. (in press-c). Understanding individual differences through developmental methods. In C. P. Benbow & D. Lubinski (Eds.), *From psychometrics to giftedness: Essays in honor of Julian C. Stanley*. Baltimore: Johns Hopkins University Press.

Keating, D. P., & Bobbitt, B. (1978). Individual and developmental differences in cognitive-processing components of mental ability. *Child Development, 49*, 155–167.

Keating, D. P., & Crane, L. L. (1990). Domain-general and domain-specific processes in proportional reasoning. *Merrill–Palmer Quarterly, 36*, 411–424.

Keating, D. P., List, J. A., & Merriman, W. E. (1985). Cognitive processing and cognitive ability: A multivariate validity investigation. *Intelligence, 9,* 149–170.

Keating, D. P., & MacLean, D. J. (1987). Cognitive processing, cognitive ability, and development: A reconsideration. In P. A. Vernon (Ed.), *Speed of information-processing and intelligence* (pp. 239–270). Norwood, NJ: Ablex.

Keating D. P., & MacLean, D. J. (1988). Reconstruction in cognitive development: A poststructuralist agenda. In P. B. Baltes, D. L. Featherman, & R. M. Lerner (Eds.), *Life-span development and behavior* (Vol. 8, pp. 283–317). Hillsdale, NJ: Erlbaum.

Kline, M. (1984). *Mathematics: The loss of certainty.* New York: Oxford University Press.

Kuhn, D., Amsel, E., & O'Loughlin, M. (1988). *The development of scientific thinking skills.* San Diego, CA: Academic Press.

Newmann, F. (in press). Higher order thinking in the teaching of social studies: Connections between theory and practice. In D. Perkins, J. Segal, & J. Voss (Eds.), *Informal reasoning and education.* Hillsdale, NJ: Erlbaum.

Shafrir, U., Ogilvie, M., & Bryson, M. (1990). Attention to errors and learning: Across-task and across-domain analysis of the post-failure reflectivity measure. *Cognitive Development, 5,* 405–425.

Stanley, J. C., Keating, D. P., & Fox, L. H. (Eds.). (1974). *Mathematical talent: Discovery, description, and development.* Baltimore: Johns Hopkins University Press.

Stevenson, H. W., & Stigler, J. W. (1991). *The learning gap.* New York: Summit Books.

Sternberg, R. J. (1985). *Beyond IQ: A triarchic theory of human intelligence.* Cambridge, England: Cambridge University Press.

Sternberg, R. J. (1989). Domain-generality versus domain-specificity: The life and impending death of a false dichotomy. *Merrill–Palmer Quarterly, 35,* 115–129.

Sternberg, R. J. (1990). *Metaphors of mind.* Cambridge, England: Cambridge University Press.

Vygotsky, L. S. (1979). *Mind in society.* Cambridge, MA: Harvard University Press.

CHAPTER 11

Adaptive and Nonadaptive Characteristics of Low-Income Children's Mathematical Strategy Use

Robert S. Siegler

Psychologists have devoted a great deal of attention to children's strategies, concepts, and rules for solving problems. Many findings in the area have considerable relevance for mathematics and science education. The "buggy" arithmetic strategies described by Brown and Burton (1978), the immature rules for solving balance scale and shadow projection problems described by Siegler (1981), and the misconceptions about physics concepts described by Kaiser, McCloskey, and Proffit (1986) all point to ways in which children's thinking deviates from the approaches that teachers are trying to inculcate. By specifying the difficulties that many children encounter and by providing means of assessing individual children's thinking, this type of research makes possible effective individualized instruction.

Funding for this research was provided by grants from the McDonnell Foundation, the Mellon Foundation, the Spencer Foundation, and the National Institutes of Health. Thanks are due to Diane Briars, director of Mathematics Education for the City of Pittsburgh, as well as to the children, teachers, and principals of Roosevelt, Chartiers, Friendship, and Bon Air elementary schools.

The examples just cited all focus on tasks in which a given child consistently uses a single rule or strategy to solve a class of problems. Recent research, however, has revealed a different case that raises a new set of issues. This case involves tasks on which children know and use multiple strategies. It now appears that such multiple strategy use is extremely common. It characterizes individual children, not just groups, and often is evident even in an individual child's efforts to solve the same problem on two occasions close in time. Arithmetic (Cooney, Swanson, & Ladd, 1988; Geary & Burlingham-Dubree, 1989; Siegler & Robinson, 1982), reading (Goldman & Saul, 1991; Jorm & Share, 1983), and spelling (Siegler, 1986; Tenney, 1980) are some of the educationally relevant areas in which children have been found to use diverse strategies.

The fact that children often use diverse strategies is not a mere idiosyncracy of their thinking. The diversity is important for both performance and learning. Strategies vary in the degree of accuracy that they yield, the amount of effort that they require for execution, the time that they take, and the range of problems to which they apply. The broader the range of strategies that children know, the more precisely they can shape their choices to the particulars of the situation and of the problem. Even young children often capitalize on the strengths of different strategies and use each one most frequently under circumstances in which its strengths are most essential and its weaknesses least damaging.

Research on strategy choice has focused predominantly on middle-income children; lower income children's strategy choices have not been the subject of much empirical research. Despite this lack of empirical study, the strategy choices of children from impoverished backgrounds have been the subject of considerable speculation. Several investigators have hypothesized that these children's poor academic performance is in large part due to poor choices of strategies, which in turn have been attributed to deficient metacognitive knowledge (e.g., Borkowski & Krause, 1983; Pressley, Borkowski, & O'Sullivan, 1984). These and similar diagnoses have led to recommendations that increased instructional emphasis be placed on teaching metacognitive skills. For example, it has been suggested that teaching children how to assess the demands of different problems would lead them to choose more effective strategies (e.g., Bransford, Sherwood, Vye, & Rieser, 1986; Paris, Saarno, & Cross, 1986).

Such diagnoses and prescriptions are in large part based on the observation that low-achieving students tend to use unsophisticated strategies. It seems at least plausible, however, that use of such strategies is primarily a consequence, rather than a cause, of the children's poor content knowledge. That is, the children may use unsophisticated strategies because they do not know how to use more sophisticated ones successfully. In this view, the performance deficiencies are not the result of choosing strategies poorly; they are the result of having only relatively slow or relatively inaccurate strategies among which to choose.

In the remainder of this chapter, I develop in greater depth this perspective on lower and middle-income children's strategy use and strategy choice. First, I summarize previous empirical findings on middle-income children's strategy choices and describe a model of how such choices are made. Then, I present recent data on the strategy choices of lower income children and on some specific features of their execution of strategies. Finally, I advance four general conclusions and a number of ideas concerning the types of instruction that will prevent small, relatively easy-to-remedy differences in early competence from developing into large, intractable, later ones.

Background

Previous studies on middle-income children's strategy choices allow us to draw at least four generalizations, discussed next.

Children Use Diverse Strategies

Most models of cognitive development postulate that children of a given age consistently use a particular strategy. This is a well-known characteristic of stage theories, but it is also true of alternative approaches. To cite one example, in the area of memory development, 5-year-olds have been described as not rehearsing, 8-year-olds as using a simple form of rehearsal, and 11-year-olds as using a more sophisticated version of the strategy (Flavell, Beach, & Chinsky, 1966; Ornstein, Naus, & Liberty, 1975). To cite another example, preschoolers have been said to add integers by counting from one, first and second graders have been said to add by counting-on from the larger addend (e.g., solving $2+4$ by counting 4, 5, 6), and older children have been said to add by retrieving the answers from memory (Ashcraft, 1982; Groen & Parkman, 1972; Ilg & Ames, 1951).

Such models have the advantage of simplicity, memorability, and ease of communication. However, careful attention to what children are actually doing has indicated that the depictions are seriously oversimplified. The first context in which I became aware of the degree of oversimplification was an experiment on 4- and 5-year-olds' addition of small numbers (Siegler & Robinson, 1982). Videotapes of the children's performance made it clear that they sometimes put up their fingers and counted them, sometimes put up their fingers and recognized how many fingers were up without counting, sometimes counted without any obvious external referent, and sometimes retrieved the answer from memory. Subsequent studies demonstrated that the variability of strategy use in single-digit addition lasts at least through third grade, with changes coming in the introduction of new backup strategies, such as counting-on and decomposition, and in increasing use of retrieval.

In the decade since then, children's use of diverse strategies has been documented in many domains. To multiply, 8- to 10-year-olds sometimes repeatedly add one of the multiplicands, sometimes write the problem and then recognize the answer, sometimes write and then count groups of hatch marks that represent the problem, and sometimes retrieve the answer from memory (Siegler, 1988a). To tell time, 7- to 9-year-olds sometimes count forward from the hour by ones or by fives, sometimes count backward from the hour by ones or fives, sometimes count from reference points such as the half hour, and sometimes retrieve the time that corresponds to the clock hands' configuration (Siegler & McGilly, 1989). To spell, 7- and 8-year-olds sometimes sound out words, sometimes look them up in dictionaries, sometimes write out alternative forms and try to recognize which is correct, and sometimes recall the spelling from memory (Siegler, 1986). Similar variability has been observed in causal reasoning (Shultz, Fisher, Pratt, & Rulf, 1986), spatial reasoning (Ohlsson, 1984), communication strategies (Kahan & Richards, 1986), and other situations.

These diverse strategies are not artifacts of one child using one strategy and a different child another one. Most individuals of all ages in all of these domains have been found to use at least two strategies; in most of the domains, the norm has been to use three or more. This multiple strategy use is apparent even within a single class of problems and, in the limiting case, on the same problem presented to the same child on consecutive days (Siegler & Shrager, 1984; Siegler & McGilly, 1989).

These experiments have called into serious question previous models in these areas postulating that children of a given age consistently use a particular strategy. To illustrate, a number of models of first and second graders' addition (Ashcraft, 1982, 1987; Groen & Parkman, 1972; Svenson, 1975) postulate that 6- and 7-year-olds consistently add by using the *min strategy* of counting-on from the larger addend (e.g., they would solve $3 + 5$ by starting with 5 and then thinking "6 is 1, 7 is 2, 8 is 3"). However, Siegler (1987b) found that children of these ages used the min strategy on only a minority of trials; on other trials, they counted from 1, retrieved answers from memory, decomposed problems into simpler form, or guessed. The earlier conclusions were found to be due to statistical artifacts that arose from averaging data over different strategies.[1] Parallel findings have arisen in studies of simple subtraction (Siegler, 1989) and of serial recall (McGilly & Siegler, 1989). In these and many other domains, there seems little question that individual children use diverse strategies.

Children Choose Adaptively Among Strategies

Using diverse strategies is helpful only if children shape their choice of strategy to the demands of the problem and of the situation. Even young children's choices have proven to be adaptive in at least three distinct ways. One way in which the choices are adaptive concerns the choice of whether to use retrieval or a backup strategy (an approach other than retrieval). The more difficult the problem, the more often children use backup strategies to solve it (see Figure 1). This relation has been found in studies of addition, subtraction, multiplication, spelling, word identi-

---

[1]For a detailed analysis of the source of these statistical artifacts, see Siegler (1987b).

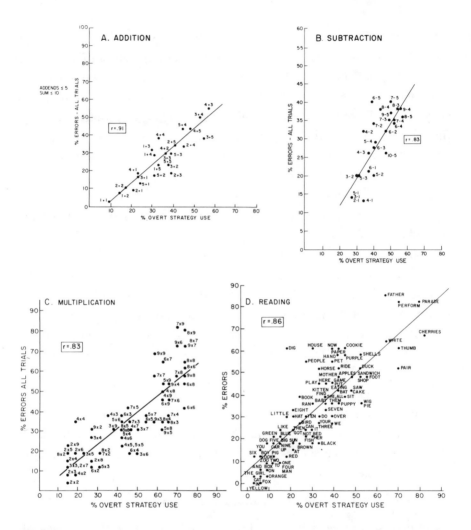

FIGURE 1. Examples of adaptive strategy choices. On all four tasks, the more difficult the problem, the more often children use backup strategies. Data from Siegler (1986).

fication, and time telling. Such choices are useful because they enable children to use the fast retrieval strategy on problems in which that approach yields correct answers and to use slower backup strategies on problems in which they are necessary to produce accurate performance. Consistent with this analysis, preventing children from using backup strategies by imposing a short time limit (4 s) produces sharp falloffs in accuracy, with the dropoffs being largest on precisely the problems on which children are most likely to use backup strategies when they are allowed to choose freely (Siegler & Robinson, 1982). Thus, the high correlations between problem difficulty and frequency of use of backup strategies are not attributable to the backup strategies being unreliable and thus causing errors on the problems. Performance on difficult problems is even worse when the backup strategies are not allowed.

A second adaptive characteristic may be seen in children's choices among alternative backup strategies. For example, in choosing whether to add by counting from one or by counting from the larger addend (the min strategy), children are especially likely to use the min strategy on problems such as 9 + 2, which have both a large difference between the addends and a relatively small minimum addend. Such problems also are the ones for which the advantage of the min strategy is greatest relative to counting from one (Siegler, 1987b).

Studies of serial recall have led to appreciation of a third sense in which children's strategy choices are adaptive: trial-to-trial changes in strategy use. Children have been found to change strategies most often when they have both erred on the previous trial and used a relatively error-prone strategy on that trial (McGilly & Siegler, 1989). When they succeeded despite using the relatively error-prone strategy, or when they had used the most accurate strategy but still erred, they were significantly less likely to switch strategies on the next trial. Again this seems adaptive; if an easy-to-execute but relatively inaccurate strategy leads to success, there is no reason to change; if the most accurate available strategy leads to failure, there is no reason to think that another approach will be more successful.

To summarize, children use multiple strategies and choose among them adaptively in a wide variety of domains. In these areas, learning and development involve a continuing competition among alternative strategies, with the faster and more accurate strategies gradually becoming dominant. Rather than one strategy being suddenly replaced by another, the competition among strategies often appears to last for several years, even on a single task such as integer addition or subtraction.

Individual Differences Also Influence Strategy Choices

Although there are substantial commonalities in children's strategy choices, there also are substantial individual differences that influence performance across domains as well as within individual areas. To learn about them, Siegler (1988b) had each of a group of first graders perform three tasks: addition, subtraction, and reading (word identification). The addition and subtraction tasks involved presentation of simple problems of the form "How much is $6 + 8$?" or "How much is $8 - 6$?" The reading task involved presenting children with a set of index cards, each with a single word on it, and asking "What does this say?" The children's accuracy, speed, and strategy use on each trial provided the main data.

A cluster analysis of the children's percent correct on retrieval trials, percent correct on backup strategy trials, and percent use of retrieval indicated considerable consistency in performance across the three tasks. The clustering program divided children into three groups, which I labeled the *good students*, the *not-so-good students*, and the *perfectionists*.

The contrast between good and not-so-good students was evident along all of the dimensions that might be expected from the names (see Figure 2). Good students used retrieval more often in both addition and subtraction. They were correct more often on both retrieval and backup strategy trials on all three tasks. They also were faster in executing backup strategies on all three tasks and faster in using retrieval on both addition and subtraction problems.

The relation of the performance of the perfectionists to that of children in the other two groups was more complex. Despite their being as fast and as accurate as the good students on both retrieval and backup strategy trials, the perfectionists used retrieval less often than even the not-so-good students. Convergent validation for these findings was obtained in a second experiment with a somewhat different strategy assessment method, different problems, and different children (Siegler, 1988b, Experiment 2).

Four months after the experiment, all children were given the Metropolitan Achievement Test. Differences between perfectionists and good students on the one hand and not-so-good students on the other echoed those in the experimental setting. As shown in Table 1, the perfectionists' average mathematics scores were at the 81st percentile, the good students' at the 80th percentile, and the not-so-good students' at the 43rd percentile. Adding further convergent validity, 44% of the not-so-good students were

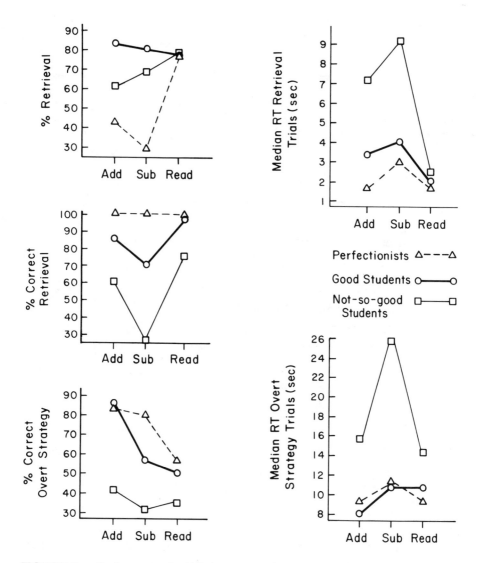

FIGURE 2. Performance of middle-income good students, not-so-good students, and perfectionists. Data from Siegler (1988b).

either held back or classified as learning disabled the next year, versus none of the good students and perfectionists. Thus, differences between not-so-good students and the other two groups would have been evident on standardized achievement tests and on school placements, whereas differences between good students and perfectionists would not have been.

TABLE 1

Achievement Test Scores (Percentiles) of Middle-Income Perfectionists, Good Students, and Not-so-Good Students

Measure	Perfectionists	Good students	Not-so-good students
Total mathematics	86	81	37
Mathematics computation	84	68	22
Mathematics problem solving	80	80	38
Total reading	81	83	52
Word recognition	79	84	54
Reading comprehension	76	83	57

The Strategy Choice Model Accounts for Both Regularities and Individual Differences in Strategy Choices

Most previous theorizing about how children choose strategies has emphasized the role of metacognitive knowledge. The dominant view has been that children assess the difficulty of problems, available strategies, and the likely success of the strategies for solving the problems, and then decide which strategy to use. This perspective has led to a great many classroom efforts to teach more advanced metacognitive skills.

Although such metacognitive skills may be important in some strategy choices, they do not seem to be the main mechanism of strategy selection in all cases. One sign of this was a finding of Siegler and Shrager (1984). They asked children whether each of a set of arithmetic problems was "easy," "hard," or "in between," and then examined how closely the children's judgments of problem difficulty paralleled their frequency of backup strategies on the same problems. Although the judgments of problem difficulty were somewhat related to how often they used backup strategies, the relation was much weaker than that between their strategy choices and the objective difficulty of the problem (measured in terms of percentage of errors or solution times). This suggested that children's choices reflected knowledge not captured in their conscious explicit judgments of problem difficulty.

If not through a rational consideration of problem difficulty and characteristics of available strategies, how would children choose which strategy to use? The basic answer is that in domains in which people have large amounts of experience, associative knowledge concerning problems, answers, and strategies may guide their choices.

To illustrate how such associative knowledge could give rise to adaptive strategy choices even in the total absence of abstract metacognitive understanding, Siegler and Shrager (1984) formulated a computer simulation of how children choose among competing strategies in single-digit addition. The same framework was later used to simulate the development of simple subtraction (Siegler, 1987a) and multiplication (Siegler, 1988a). The simulations not only illustrate how strategy choices are made at any one time, but also how learning occurs and how the strategy choices change with experience.

Two ideas are central within these simulations: *associative strength* and the *confidence criterion*. Associative strength refers to the links between problems and potential answers to them. Probability of retrieving the correct answer on a given problem is a function of

$$\frac{\text{Associative strength of the correct answer}}{\text{Summed associative strengths of all answers}} .$$

The confidence criterion is a threshold that a retrieved answer must exceed for it to be stated. If the associative strength of the answer that is retrieved exceeds the confidence criterion, that answer is stated. If not, the child uses a backup strategy to solve the problem.

The model's basic workings can be described quite simply. At the beginning of each run of the simulation, the program almost always solves problems by executing backup strategies, such as counting fingers. These often, but not always, allow the program to generate the right answer.[2]

Each time an answer for a problem is generated, the association between the problem and that answer increases. This leads to the correct answer gaining increasing associative strength, and the strength exceeding more and more confidence criteria. This allows the retrieved answer to be stated on an increasing percentage of trials. Because no single

---

[2]To illustrate how wrong answers are generated in the simulation, when children use the counting fingers strategy, there is a certain probability of error on each count generated either by double counting or by skipping objects in the representation. This leads to more errors occurring on problems with larger addends, in which the representation must include more objects.

incorrect answer is frequently generated by the backup strategies, none of them gains much associative strength; thus, when a retrieved answer can be stated, it is generally correct. The simulation is more complicated than this simple depiction suggests, but the depiction does outline the basic elements and the processes that lead to increasing use of retrieval and increasingly correct performance with experience.

Results of the simulations indicated that these mechanisms lead to performance that closely parallels that of children solving the same problems. For example, in multiplication, the correlations between the children's and the simulation's percent correct on each problem, percent backup strategy use on each problem, and mean solution time on each problem were .90, .90, and .95. Thus, the same problems were the most difficult for the simulation as for the children, and the simulation, like the children, chose to use backup strategies most often on those problems. Also like children, the simulation showed substantial learning from its experience with problems. Percent correct increased from 61% during the first 10% of the problems to 89% on the last 10%. Over the same period, percent use of retrieval increased from 24% to 77% of trials.

Recently, a new version of the simulation (Siegler & Shipley, in press) has proved capable not only of generating these relations but also of generating the same patterns of individual differences in strategy choices seen in children. The differences were produced by varying two parameters (the accuracy with which backup strategies are executed and the stringency of confidence criteria) and leaving the rest of the simulation constant. Having the simulation execute backup strategies very accurately and setting fairly high confidence criteria led to performance like that of the good students; having the simulation execute backup strategies inaccurately and setting low confidence criteria led to performance like that of the not-so-good students; having the simulation execute backup strategies very accurately and setting very high confidence criteria led to performance like that of the perfectionists. These results demonstrate that the proposed mechanisms are sufficient to account for the observed individual differences as well as for the general characteristics of strategy choice.

Low-Income Children's Strategy Choices

Very recently, Dennis Kerkman and I examined whether children from impoverished backgrounds would show similar patterns of strategy use and strategy choice and similar individual differences in these properties to those observed with middle-income children (Kerkman & Siegler, 1993). The study paralleled exactly the Siegler (1988b) study of individual differences, except that the children were examined somewhat later in first grade (April rather than November and December). The reason for the later testing was to achieve greater similarity in absolute levels of performance; the goal was to examine patterns of strategy choices and of individual differences, equated as closely as possible for absolute levels of performance.

Within the low socioeconomic status sample, 80% of children came from sufficiently impoverished homes that they qualified for the federally subsidized lunch program; 71% were African–American. The addition, subtraction, and reading items presented to the children were largely identical to those used in the earlier study of individual differences.

General Level of Performance

As hoped, absolute levels of accuracy, speed, and strategy use of the lower income sample were comparable to those of middle-income children tested 5 months earlier in the school year. The middle-income sample in Siegler (1988b) was somewhat more accurate in addition (84% vs. 74%) and subtraction (70% vs. 55%), and the lower income sample in Kerkman and Siegler (1993) was more accurate in word identification (86% vs. 54%). Similarly, the middle-income children retrieved answers somewhat more often on addition problems (49% vs. 43%), the two groups retrieved equally often on subtraction problems (35%), and the lower income children retrieved more often on the word identification task (75% vs. 52%).

Adaptiveness of Strategy Choices

The lower income children also showed highly adaptive choices among alternative strategies. This may be seen in their choices of retrieval or backup strategies on different problems. As shown in Table 2, their strat-

TABLE 2

Adaptiveness of Strategy Choices of First Graders From Lower and Middle-Income Families

Group	Addition	Subtraction	Word identification
Lower income children	.91	.92	.86
Middle-income children	.87	.83	.84

Note. Correlations are between percent backup strategy use and problem difficulty (percent errors).

egy choices were just as closely related to the problem characteristics as were the choices of those in the middle-income sample. Percent backup strategy use was also highly correlated with percent errors on retrieval trials for addition ($r = .85$), subtraction ($r = .86$), and word identification ($r = .77$). Like middle-income children, children from impoverished backgrounds were much more likely to use backup strategies on problems in which they were not likely to retrieve the correct answer.

Individual Differences

Initial Classifications

As in the studies of middle-income children, cluster analyses were performed to examine patterns of individual differences. The analyses again identified three distinct groups that fit the descriptions of good students, not-so-good students, and perfectionists (see Figure 3). As with the middle-income sample, the good students were more accurate than the not-so-good student group the least stable, and those in the perfectionist group in between (see Table 4). The changes seemed to involve mainly a shift toward the good student pattern of accurate performance using both when they retrieved and when they used backup strategies, but they used retrieval even less often than the not-so-good students. Once again, achievement test performance (scores on the California Achievement Test) provided convergent validation for the difference between the perfectionists and the good students on one hand and the not-so-good students on the other. Performance of perfectionists and of good students

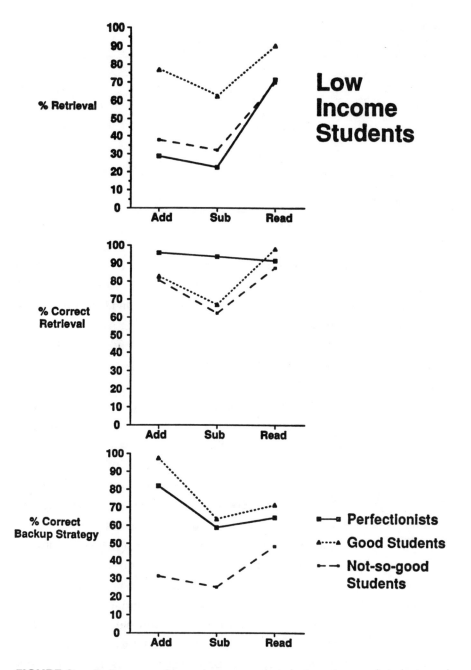

FIGURE 3. Performance of lower income good students, not-so-good students, and perfectionists. Data from Kerkman and Siegler (1993).

TABLE 3
Achievement Test Scores (Percentiles) of Low-Income Children

Measure	Perfectionists	Good students	Not-so-good students
Reading comprehension	59	65	36
Language expression	55	69	33
Vocabulary	64	69	44
Word analysis	57	75	50
Total reading	63	69	39
Mathematics computation	59	64	44
Mathematics applications	65	67	40
Total mathematics	64	70	43

did not differ significantly on any of the tests, and both groups were consistently superior to the not-so-good students (see Table 3).

The proportions of children in the three individual-differences groups also were similar to those in the middle-income sample: 43% perfectionists, 23% good students, and 34% not-so-good students in the low-income sample versus 41% perfectionists, 35% good students, and 24% not-so-good students in the middle-income sample. An additional noteworthy characteristic of the composition of the three groups in both lower and middle-income samples was the lack of gender differences. The percentages of boys and girls within each group closely paralleled those for the samples as a whole. Thus, although the description of the perfectionists fit a stereotype associated with young girls and the description of the not-so-good students that of a stereotype associated with young boys, the data indicate that the patterns are independent of gender.

The adaptive quality of strategy choice in the lower income sample extended to all three individual-differences groups. Even the lower income not-so-good students made highly adaptive choices. The correlations between each problem's percent backup strategy use and its percent errors were .87, .89, and .77 for addition, subtraction, and word identification, respectively. It appears that on tasks such as arithmetic and reading, with which students have considerable experience, even low-achieving children within low-income populations choose strategies in entirely reasonable ways. Whatever their abstract metacognitive knowledge about their own abilities, the problems, and the strategies, they are able to choose whether to use a backup strategy quite effectively.

Longitudinal Stability

Kerkman and Siegler (1993) also examined stability of the individual-differences groups over time. We returned to the same school when the children were in second grade and retested the children who remained (35 of the 41 from the original sample). The sample was more uniformly lower in income than it had been in first grade; now, 97% of children qualified for the federally subsidized lunch program. As previously, 71% of the children were African–American.

Despite a number of procedural differences between the first and second testings (most important, only the addition task was administered on the second occasion), 63% of children were classified as being in the same individual-differences group. The small number of subjects in each group limits the confidence with which generalizations on the relative stability of the different groups can be drawn. However, classifications in the good student group seemed to be the most stable, those in the not-so-good student group the least stable, and those in the perfectionist group in between (see Table 4). The changes seemed to involve mainly a shift toward the good student pattern of accurate performance using both retrieval and backup strategies and high reliance on retrieval. This is the goal of instruction in these areas; students do move toward this goal over time.

The longitudinal design also allowed us to test a central prediction of the model that had never previously been tested, that early accuracy

TABLE 4

Number of Children in Each Individual-Differences Group in First and Second Grades

	Classification in second grade		
Group	Perfectionists	Good students	Not-so-good students
Classification in first grade			
Perfectionists	10	5	0
Good students	0	8	0
Not-so-good students	4	4	4

of performance using backup strategies should correlate negatively with later frequency of use of backup strategies. This paradoxical-sounding prediction comes about for a straightforward reason: Children who execute backup strategies accurately will develop the strongest associations between problems and their correct answers, and the weakest associations between problems and their wrong answers. This will lead to the correct answer being retrieved more often, and having enough associative strength more often to exceed the confidence criterion and therefore be stated.

The data were consistent with this prediction. Percent correct on addition backup strategy trials in first grade was significantly and negatively correlated with percent use of addition backup strategies in second grade (addition was the only task presented in both grades). As also would be expected from the model, percent correct on addition backup strategy trials in first grade was significantly and positively correlated with percent correct on retrieval trials in second grade. This was not attributable to all three variables being correlated with percent retrieval use in first grade. Even after partialing out percent retrieval in first grade, both correlations remained significant. Thus, accurate execution of backup strategies appears to help children learn to retrieve accurately, which allows them to decrease their use of backup strategies and to increase their use of retrieval.

Strategy Execution

In a very recent study (Siegler & Kerkman, 1993), we focused on the details of strategy execution of 113 first graders from lower income backgrounds. Each subject was presented 100 addition problems with addends 0–9. Strategies were assessed through the same combination of videotapes of overt behavior and immediately retrospective verbal reports as in Siegler (1987b). This very large data set of 11,300 separate assessments of strategies allowed a detailed picture of the use of both common and uncommon strategies.

The general pattern of strategy use resembled closely that which had emerged previously. On more than 90% of trials, children used strategies that had been observed often in previous studies. In order of frequency of use, these strategies were retrieval (51% of trials), counting from the larger addend (16%), counting from one (14%), finger recognition (7%), guessing (3%), and saying "I don't know" (3%).

On the remaining 5% of trials, however, the children used two other strategies. One was an approach that we had rarely observed in our previous detailed strategy assessment studies (all of which had been conducted with middle-class children); the other strategy of interest was one that we literally never had observed before. The very large data set meant that, even though these two strategies were relatively infrequent, we still observed them on a large number of trials (415 in one case, 31 in the other). Thus, we could be confident that they were being used; their existence did not depend on assessments from a small number of trials.

The more common of the two approaches was counting-on from the smaller addend. On a problem such as 5 + 9 or 9 + 6, the child would count-on from the 5. The less common of the approaches was counting on from neither addend. On 5 + 9 or 9 + 5, the child might count on from 7 or 8.

First, consider the strategy of counting-on from the smaller addend. To determine the types of problems on which children used this strategy, we performed a regression analysis of the 2,104 trials on which children counted from either the smaller or the larger addend. The dependent variable was each problem's percentage of trials on which the child counted from the smaller (rather than from the larger) addend.

By far the best predictor of how often children counted from the smaller addend was whether it was also the first addend. Children counted far more often from the smaller addend when it came first than when it came second (more often on 5 + 9 than on 9 + 5). This variable accounted for 64% of the variance in frequency of counting from the first addend. The other significant predictor was the size of the smaller addend. Children counted more often from the smaller addend when it was larger than when it was smaller. Thus, they counted more often from 5 on 5 + 9 than from 2 on 2 + 9. The two variables together accounted for 75% of variance in the percentage of trials on each problem on which children counted from the smaller addend.

Analyses of individual children's counting from the smaller addend were especially revealing. One group (36% of children) counted from the smaller addend on fewer than 10% of the trials on which they counted from one of the two addends. Another group (11% of children) counted from the smaller addend on at least 50% of the trials on which they counted from one or from the other addend. However, the largest group (53%) counted from the smaller addend on 10% to 50% of such trials. These data demonstrate that counting from the smaller addend could not be attributed to a small number of children engaging consistently in the activity. Rather, it seemed to be something that many of the lower income children did sometimes. The pattern contrasted with the findings of Siegler (1987b) and of Siegler and Jenkins (1989), studies that applied the same strategy assessment methodology to middle-income samples. In both of the studies with middle-income samples, counting from the smaller addend was an extremely rare activity. Only one of the 4- and 5-year-olds in the Siegler and Jenkins study and none of the middle-income 6- and 7-year-olds in Siegler (1987b) were observed to use it versus 75% of the children from the low-income families in Siegler and Kerkman (1993).

The other strategy of interest was counting from neither addend. My colleagues and I had never observed this approach in any of the previous studies of children from middle-income backgrounds.[3] Even in Siegler and Jenkins (1989), where we observed 4- and 5-year-olds' very first uses of counting from a number larger than one, the children never counted from a number different from either addend.

In the Siegler and Kerkman (1993) study, 10 children (9% of the sample) sometimes counted from neither addend. They did this on 20% of the trials on which they counted from some number other than one. In most instances (96%), such counts occurred on large addend problems—problems in which the sum was 11 or more. This was a substantially higher percentage than would have been expected from the percentage of such problems in the total set (59%).

---

[3]The one previous study that I conducted with lower income children (Kerkman & Siegler, 1993) divided strategy use only into retrieval strategies and backup strategies. This method did not allow discrimination among the different backup strategies that those lower income children used.

What, if anything, does counting from the smaller addend or counting from neither addend signify? Although counting from the smaller addend sometimes generates correct answers, and counting from neither addend hardly ever does, both strategies may reflect different degrees of a similar problem: an early gap between these children's use of a strategy and their understanding of why the strategy works and of what goals it achieves. Counting from the smaller addend has the same disadvantage as counting from the larger addend—it demands that children count from an unfamiliar starting point—without having as great an advantage in reducing the amount of counting as is needed. The children who were observed using these strategies were in the somewhat atypical situation of being in a district in which teachers explicitly taught the strategy of counting from the larger addend. This instruction may have led them to count from an addend without understanding why it was more desirable to count from the larger than from the smaller one. Without understanding why the strategy was both legitimate and effective, the children at times dropped even the requirement to count from an addend, thus giving rise to counting from neither addend.

All this is speculation. However, it does have two implications. First, it suggests that the often-noted separation between procedural and conceptual understanding (Hiebert, 1986) may begin quite early, at least for some children. Second, it suggests a need to assess children's conceptual understanding of their strategies and of why the strategies can be used as well as to assess which strategies they use and when they use them.

Conclusion

These findings suggest several general conclusions. First, there seems little question that on tasks in which they have experience, children from low-income families choose strategies very adaptively. This was evident in the very high correlations on addition, subtraction, and word identification between each problem's difficulty and how often children used a backup strategy to solve it. The pattern held for all three individual-differences groups. Even lower income not-so-good students chose to use such strategies in situations in which they would do the most good.

Whatever the children's abstract, explicit metacognitive knowledge, their associative knowledge allows them to make at least one strategy choice very adaptively.

Second, children from low-income families exhibit the same types of individual differences as children from middle-class backgrounds. Both groups exhibit the good student, not-so-good student, and perfectionist patterns. Furthermore, the percentage of children fitting each pattern is quite similar within the two groups. The classifications predict achievement test performance in the sense that both good students and perfectionists consistently outperform the not-so-good students. However, the strategy classifications allow us to go beyond the achievement test scores to differentiate between two groups of successful students.

Third, the strategy choice model accounts for both general patterns of performance and for individual differences within the middle- and lower income samples. The simulation demonstrated that the underlying model was sufficient to produce the same increases in accuracy, the same shifts in strategy use from primary reliance on backup strategies to primary reliance on retrieval, the same adaptive strategy choices, and the same types of individual differences as observed in the children. The model also pointed to two primary sources of these general patterns and individual differences: success in executing backup strategies, which contributes to the associative strengths linking problems and answers, and the confidence criteria, which determine when a retrieved answer will be stated. The model also indicated how a child lacking metacognitive knowledge could nonetheless choose reasonably among alternative strategies.

The fourth and final conclusion is that taking into account early individual differences in strategy use, strategy choice, and understanding of strategies may lead to better instruction. Our findings have several specific implications concerning how this can be accomplished.

The first approach considers the group most in need of special instruction, the not-so-good students. Our results indicate that they need help in executing backup strategies more effectively. At present, many teachers discourage such children from using such backup strategies. When I interviewed 6 first-grade teachers about the practice, they expressed a belief that backup strategies are a crutch that prevents children from retrieving answers effectively. The reasoning seemed to be that older and higher achieving children usually use retrieval, that younger and lower achieving children usually use backup strategies, and therefore that the way to make younger and lower achieving children more like older and higher achieving peers is to forbid their use of backup strategies.

In contrast, the model and results described in this chapter suggest that adopting such an attitude is akin to blaming the messenger for the bad news. When children know correct answers sufficiently well to retrieve them, they do so spontaneously. Preventing them from using backup strategies when they do not know the answers leads to a great deal of incorrect performance. It is hard to see how this approach can aid learning.

A far more promising instructional approach would be to help children execute backup strategies more effectively. Children who execute backup strategies accurately in first grade actually use them less in second grade because they no longer need them as much, because past successful use of the backup strategies has allowed them to learn to retrieve the correct answer. Thus, teaching children early on to execute backup strategies accurately may lead them to spontaneously stop using them later.

Another approach that may work well with not-so-good students is to help them understand why backup strategies work and why it is legitimate to use them. The lower income children's fairly frequent counting from the smaller addend and occasional counting from neither addend suggests that they did not entirely understand why counting from a number other than one was a good idea. Discussing both why counting from the larger addend is desirable and why it is legitimate may help bridge the gap between the children's early strategy use and their conceptual understanding of the strategies.

One other way of helping the not-so-good students involves persuading them to raise their confidence criteria. These children appeared to adopt very low confidence criteria for stating retrieved answers. This could be inferred from their using retrieval on a relatively high percentage of trials despite not retrieving very accurately. Such low confidence criteria may have resulted from the poor success that these children had experienced previously with backup strategies: If one is going to be wrong in any case, why not answer quickly and be done with it? From this line of reasoning, improving the children's accuracy in using backup strategies may automatically lead to their increasing their confidence criteria. It is also possible, however, that the confidence criteria will stay low even after the original reason for them being low is no longer present. If this occurs, it may be useful to tell not-so-good students to state answers from memory only when they are pretty sure the answers are correct.

The perfectionists present quite a different challenge. They are basically doing well, as indicated both by their very accurate performance in the experiment and by their high achievement test scores. However, their heavy reliance on backup strategies may slow their performance unnecessarily. When they do retrieve, they do so extremely accurately; they simply do not use this approach very often. With such perfectionists, it may be advisable to suggest that they state the answer that they think is correct, even if they are not absolutely sure. However, the impact of the suggestion should be monitored carefully. If their accuracy falls off following the suggestion, they should be encouraged to return to their previous approach. After all, they are doing well already; it just may be possible to help them do even better.

Finally, with the good students, encouragement and an occasional pat on the back are all that seem to be necessary. Their quick and accurate performance and frequent use of retrieval closely approximate the goals of instruction. With some students, standard instructional methods are effective; thus, there is no good reason to change them.

References

Ashcraft, M. H. (1982). The development of mental arithmetic: A chronometric approach. *Developmental Review, 2*, 213–236.

Ashcraft, M. H. (1987). Children's knowledge of simple arithmetic: A developmental model and simulation. In J. Bisanz, C. J. Brainerd, & R. Kail (Eds.), *Formal methods in developmental psychology* (pp. 302–338). New York: Springer-Verlag.

Borkowski, J. M., & Krause, A. (1983). Racial differences in intelligence: The importance of the executive system. *Intelligence, 7*, 379–395.

Bransford, J., Sherwood, R., Vye, N., & Rieser, J. (1986). Teaching thinking and problem-solving: Research foundations. *American Psychologist, 41*, 1078–1089.

Brown, J. S., & Burton, R. B. (1978). Diagnostic models for procedural bugs in basic mathematical skills. *Cognitive Science, 2*, 155–192.

Cooney, J. B., Swanson, H. L., & Ladd, S. F. (1988). Acquisition of mental multiplication skill: Evidence for the transition between counting and retrieval strategies. *Cognition and Instruction, 5,* 323–345.

Flavell, J. H., Beach, D. R., & Chinsky, J. M. (1966). Spontaneous verbal rehearsal in a memory task as a function of age. *Child Development, 37,* 283–299.

Geary, D. C., & Burlingham-Dubree, M. (1989). External validation of the strategy choice model for addition. *Journal of Experimental Child Psychology, 47,* 175–192.

Goldman, S. R., & Saul, E. U. (1991). Flexibility in text processing: A strategy competition model. *Learning and Individual Differences, 2,* 181–219.

Groen, G. J., & Parkman, J. M. (1972). A chronometric analysis of simple addition. *Psychological Review, 79,* 329–343.

Hiebert, J. (1986). *Conceptual and procedural knowledge: The case of mathematics.* Hillsdale, NJ: Erlbaum.

Ilg, F., & Ames, L. B. (1951). Developmental trends in arithmetic. *Journal of Genetic Psychology, 79,* 3–28.

Jorm, A. F., & Share, D. L. (1983). Phonological recoding and reading acquisition. *Applied Psycholinguistics, 4,* 103–147.

Kahan, L. D., & Richards, D. D. (1986). The effects of context on children's referential communication strategies. *Child Development, 57,* 1130–1141.

Kaiser, M. K., McCloskey, M., & Proffit, D. R. (1986). Development of intuitive theories of motion: Curvilinear motion in the absence of external forces. *Developmental Psychology, 22,* 67–71.

Kerkman, D. D., & Siegler, R. S. (1993). Individual differences and adaptive flexibility in lower-income children's strategy choices. *Learning and Individual Differences, 5,* 113–135.

McGilly, K., & Siegler, R. S. (1989). How children choose among serial recall strategies. *Child Development, 60,* 172–182.

Ohlsson, S. (1984). Induced strategy shifts in spatial reasoning. *Acta Psychologica, 57,* 47–67.

Ornstein, P. A., Naus, M. J., & Liberty, C. (1975). Rehearsal and organizational processes in children's memory. *Child Development, 26,* 818–830.

Paris, S. G., Saarno, D. A., & Cross, D. R. (1986). A metacognitive curriculum to promote children's reading and learning. *Australian Journal of Psychology, 38,* 107–123.

Pressley, M., Borkowski, J. M., & O'Sullivan, J. T. (1984). Memory strategy instruction is made of this: Metamemory and durable strategy use. *Educational Psychologist, 19,* 94–107.

Shultz, T. R., Fisher, G. W., Pratt, C. D., & Rulf, S. (1986). Selection of causal rules. *Child Development, 57,* 143–152.

Siegler, R. S. (1981). Developmental sequences within and between concepts. *Monographs of the Society for Research in Child Development, 46* (Whole No. 189).

Siegler, R. S. (1986). Unities in strategy choices across domains. In M. Perlmutter (Ed.),

Minnesota Symposium on Child Development (Vol. 19, pp. 1–48). Hillsdale, NJ: Erlbaum.

Siegler, R. S. (1987a). Strategy choices in subtraction. In J. Sloboda & D. Rogers (Eds.), *Cognitive process in mathematics*. Oxford, England: Oxford University Press.

Siegler, R. S. (1987b). The perils of averaging data over strategies: An example from children's addition. *Journal of Experimental Psychology: General, 116*, 250–264.

Siegler, R. S. (1988a). Strategy choice procedures and the development of multiplication skill. *Journal of Experimental Psychology: General, 117*, 258–275.

Siegler, R. S. (1988b). Individual differences in strategy choices: Good students, not-so-good students, and perfectionists. *Child Development, 59*, 833–851.

Siegler, R. S. (1989). Hazards of mental chronometry: An example from children's subtraction. *Journal of Educational Psychology, 81*, 497–506.

Siegler, R. S., & Jenkins, E. (1989). *How children discover new strategies*. Hillsdale, NJ: Erlbaum.

Siegler, R. S., & Kerkman, D. D. *Atypical strategy use by lower income children*. Manuscript in preparation.

Siegler, R. S., & McGilly, K. (1989). Strategy choices in children's time-telling. In I. Levin & D. Zakay (Eds.), *Time and human cognition: A life span perspective* (pp. 185–218). Amsterdam: Elsevier.

Siegler, R. S., & Robinson, M. (1982). The development of numerical understandings. In H. W. Reese & L. P. Lipsitt (Eds.), *Advances in child development and behavior* (pp. 241–312). San Diego, CA: Academic Press.

Siegler, R. S., & Shipley, C. (in press). Variation, selection, and cognitive change. In G. Halford & T. Simon (Eds.), *Developing cognitive competence: New approaches to process modeling*. Hillsdale, NJ: Erlbaum.

Siegler, R. S., & Shrager, J. (1984). Strategy choices in addition and subtraction: How do children know what to do? In C. Sophian (Ed.), *Origins of cognitive skills* (pp. 229–243). Hillsdale, NJ: Erlbaum.

Svenson, O. (1975). Analysis of time required by children for simple additions. *Acta Psychologica, 39*, 289–302.

Tenney, Y. J. (1980). Visual factors in spelling. In U. Frith (Ed.), *Cognitive processes in spelling* (pp. 215–229). San Diego, CA: Academic Press.

Index

Research (continued)
psychological, educational reform
and, 15–16, 153–155
scientist–practitioner model in
education, 277–278
Research needs
on development of mathematical
reasoning, 268–269, 270
on developmental source of
individual differences, 327–329
on gifted students, 41–42
on information processing models,
180–187
on integration of cognitive activities,
324
in learner-centered education,
309–311
in very-long-term memory, 71–72
Residential schools, 31
Resource allocation theory, 55, 88–91, 95
Role models, female, 37–39, 40, 246–247

Schema theory, 92
Scholastic Aptitude Test. *See also*
Standardized tests
demographics, 23
gender differences in, 36, 254, 255
minority performance on, 3–4
sociocultural factors in performance
on, 23
School Development Program, 224–225
Secondary school
current levels of performance in, 2, 21
gender differences in, 241–242, 243
international comparisons, 2–3
long-term retention of course content,
71
Self-affirmation theory, 209–210
Self-view
disidentification in, 209–210, 225–226

efficacy training program, 222–223
encouraging self-confidence, 363–364
gender stereotyping in mathematics,
244–245, 248
internalized inferiority, 211–212
learning practices and, 160–161
mathematics self-confidence, 245–246,
248
personal epistemologies about
learning, 161–162, 219
social context effects in school,
227–229
test anxiety and, 103
Sex discrimination/sexual harassment,
247, 253–256
Sex-segregated classrooms, 37
Situated learning, 218–220
Situativity theory, 153–155
Social interaction
affordances/constraints in theory of,
166–167
in communities of practice concept,
156–157
in constructivist approach to learning,
57, 58
conversation analysis, 184–185
effects in learning, 221
as factor in individual differences,
330–331
in learner-centered instruction,
296–297
legitimate peripheral participation as,
156–157
model of intellectual development
and, 325–327
in research on learning, 155–159
role of concepts in, 167–168
situativity theory, 153–155
in theory of implicit understanding,
59–60

About the Editors

L ouis A. Penner received his PhD in Social Psychology from Michigan State University in 1969. He is currently a professor in the Psychology Department at the University of South Florida and was chair of the department for seven years. He is a consulting editor for the *Journal of Social and Clinical Psychology* and a member of the APA Committee on Accreditation.

George M. Batsche received his EdD in School Psychology from Ball State University in 1978. He is currently a professor in the School Psychology Program in the Department of Psychological and Social Foundations at the University of South Florida. He is a Fellow of Division 16 of APA and the immediate past president of the National Association of School Psychologists.

Howard M. Knoff received his PhD in School Psychology from Syracuse University in 1980. He is currently professor in and director of the School Psychology Program in the Department of Psychological and Social Foundations at the University of South Florida. He is a fellow of Division 16 of APA and a past president of the National Association of School Psychologists.

Douglas L. Nelson received his PhD in Experimental Psychology from the University of Wisconsin (Madison) in 1967. He is currently professor in and director of the Experimental Psychology Program in the Psychology Department at the University of South Florida. He is a fellow of Division 3 of APA and a consulting editor for the *Journal of Experimental Psychology: Memory and Cognition.*